Classic Heathkit Electronic Test Equipment

By Jeff Tranter

First Printing: November 2013.

Printed in The United States of America.

ISBN: 978-0-9921382-0-2

This book was developed using Open Source/Free software including the LibreOffice office suite, GIMP image manipulation program, and the Linux-based Ubuntu desktop operating system.

This book is dedicated to Warner Williams, a mentor who encouraged my early interest in electronics by generously giving me several pieces of test equipment including a Heathkit IM-13 VTVM.

Table of Contents

Preface

My introduction to Heathkit was in 1974 when my father and I shared a Christmas gift of a PT-15 darkroom timer kit. We built the kit together following the instructions in the assembly manual. Heathkit was renowned for the high quality of their manuals (their motto was "We won't let you fail"), and it was quite straightforward to put together. However, testing of the assembled unit indicated that it didn't work quite right. I made some voltage measurements with my trusty Radio Shack Volt-Ohm-Milliammeter (VOM) and we made a trip to the Mississauga, Ontario Heathkit store. The service technician at the store looked at the symptoms and voltage measurements I had made and suspected it was a faulty Silicon Controlled Rectifier (SCR) and gave us a new part at no cost. Sure enough, after soldering in the new part it worked flawlessly. We used the timer for a number of years in our home darkroom. A few years later when I became interested in amateur radio my first receiver was a used Heathkit HR-10B, and when I got my license I bought a used DX-60B transmitter. A Christmas gift the

Heathkit Electronic Darkroom Timer
A "must" for every well-equipped darkroom—a true electronic timer with no moving parts to reduce reliability or accuracy. The PT-15 provides accurate exposure for all contact printers and enlargers with two switch-selected ranges — up to 99 seconds in 1 second steps or up to 9.9 seconds in 0.1 second steps. At time/focus switch turns the safelight off when focusing and printing. Repeatable exposures with 2% accuracy.
Kit PT-15, Shpg. wt. 4 lbs. 36.95

$36.95

same year was an HG-10 VFO to work with the transmitter. I made regular trips to the Heathkit store to see what new products they had, including the computer system kits they started to offer in the early 1980s. Most of the equipment, like an oscilloscope, was beyond the budget of a high school student working part-time at four dollars an hour.

Flash forward a number of years, and with children raised, the advent of the Internet and eBay, and more disposable income, I reawakened my interest in Heathkit equipment and started acquiring a small collection of Heathkit shortwave, ham radio and test equipment, including the Heathkit oscilloscope that I had my eyes on all those years ago.

Illustration 1: My Ham Radio Station in 1977

Illustration 2: The Same Equipment in 2013

Acknowledgements

The inspiration for this project came from two other books. Chuck Penson's *Heathkit: A Guide to the Amateur Radio Products* has been a source of many hours of enjoyable reading and looking at pictures of Heathkit's line of ham radio equipment. It also was the impetus that drove me me purchase a number of Heathkit ham radio and shortwave equipment from eBay and other sources. My hope is for this book to attempt to do, in some small way, for Heathkit's test equipment line what Chuck's book did for their amateur radio products.

Tube Testers and Classic Electronic Test Gear, by Alan Douglas, was another inspiration which convinced me that a book on test equipment could be of interest to others. His book covers Heathkit tube testers in much more detail than I have here.

Much of the material from this book was researched and obtained from various Internet web sites too numerous to mention, but I would like to single out one site that was particularly helpful: nostalgickitscentral.com. The list of Heathkit test equipment there was the starting point for the list presented in Appendix B. Some of the images in this book came from old Heathkit catalogs I own, while photos are of equipment from my personal collection.

Thanks go to the technical reviewers of this book, Alan Douglas and Hans Gatu, who made many helpful suggestions and pointed out some embarrassing technical errors. I would also like to thank my wife Veronica for helping with copy editing and proofreading.

Finally, I encourage you, the reader, to send me any corrections or feedback in general at the e-mail address listed below. I'm also interested in hearing if there is interest in a e-book version, and if so, what formats you would like to see it offered in.

Jeff Tranter <tranter@pobox.com>
Ottawa, Ontario, Canada
November, 2013

Other than test equipment, Heathkit sold products in the following major categories:

- Amateur Radio

- Audio/Stereo/High Fidelity

- Automotive

- Citizen's Band Radios

- Clocks

- Computers

- Educational Courses

- Home Products

- Marine

- Miscellaneous/General Products

- Music

- Photography

- R/C Modeling

- Security

- Shortwave Listening

- Television

- Tools

- Weather Monitors

Structure of this Book

In this book I attempt to cover the majority of electronic test equipment that was offered by Heathkit through the years. I don't include in this category equipment such as meters and power supplies that were specific to amateur radio, equipment for automotive use like ignition analyzers, or equipment that was more of a scientific nature like pH meters and chart recorders. I do include some miscellaneous equipment that are not test instruments as such but are interesting because they served some unique purpose, like Heathkit's early analog computers. I also don't cover in detail the various electronic educational courses that were offered by Heathkit, sometimes in conjunction with hardware.

Chapter 1 gives a brief history of Heathkit, from its early beginning selling aircraft kits, to their most recent attempt to get back into the kit business. Chapter 2 gives an overview of Heathkit's lines of test equipment, including the various product numbering systems that were used. Chapter 3 provides some tips on buying vintage test equipment from sources like eBay and flea markets and restoring it to working order.

Chapters 4 through 11 cover the test equipment models, broken down into major categories: Component Testers and Substitution Boxes, Frequency Counters, Meters, Oscilloscopes, Power Supplies, Signal Generators, Tube Testers and Checkers, and finally Miscellaneous Test Equipment. The categories are somewhat arbitrary and are of my own choosing. Each chapter concludes with one or more "In-Depth" sections that look at a representative model from my Heathkit collection covering its features, operation, and any quirks or trivia about it as well as details on the unit I own and its restoration.

Appendix A offers a list of references and resources including books, web sites, and suppliers of parts, manuals, and related products and services. Appendix B contains a Product Listing that attempts to list all known test equipment produced by Heathkit, including some not covered in the book. Finally, the Index gives an alphabetical listing of the topics and product models covered in the book.

Some interesting Heathkit facts and trivia:

- Apple co-founder Steve Jobs built some Heathkits as a teenager. The other co-founder, Steve Wozniak, was a ham radio operator and may have owned some Heathkits as well, or would have at least been familiar with them.

- Heathkit offered one of the first consumer robot kits, the HERO 1. Introduced in 1982, it featured a 6808 microprocessor with four kilobytes of memory and sensors for light, sound, and motion. Programs could be stored on cassette tape and a speech synthesizer and robotic arm were available as options. It is now a sought after collector's item.

- In the 1960s, Heathkit offered a line of electric guitars under the *Harmony* brand name, as well as guitar amplifier kits.

- Heathkit pioneered motion sensor lamps. The lighting and security business was spun off as a separate company called *HeathCo* and continues to sell products under the *Heath/Zenith Lighting Controls* brand.

- It is estimated that Heathkit sold more than two million VTVMs over the period they were offered from the late 1940s to early 1990s.

- The 2000 Hollywood science fiction film "Frequency" features a character who is able to contact his deceased father 30 years in the past using a ham radio. The radio used in the film was a Heathkit SB-301. Observant fans have pointed out that the SB-301 is a receiver only, and is not capable of transmitting.

- The GC-1 Mohican was the first fully transistorized shortwave radio receiver offered by Heathkit, and is believed to be the first commercial all solid-state shortwave receiver on the market.

- Over the years, Heathkit offered various credit and financing plans for their kits so that purchasers could pay by monthly installments.

- Howard Nurse built many Heathkits as a boy, beginning with a DX-40 ham radio transmitter. He eventually become president of the company, serving in the position from 1966 to 1980, and was said to have built hundreds of Heathkits.

- While the company was not publicly traded and did not release financial figures, it is estimated that in the 1970s Heathkit had annual sales of about 100 million dollars.

- Heathkit color televisions were some of the most complex, desirable, and expensive kits. In 1983 the top of the line GR-3000 25" color television kit sold for $1099.95, equivalent to about $2500 in 2013 dollars.

- US senator Barry Goldwater was a radio amateur and built more than one hundred Heathkits. He used to fly to Heath's headquarters twice a year in his private plane to buy kits.

Chapter 1: A Brief History of Heathkit

The Heath Company was founded by Edward Bayard Heath in 1912 as an aircraft company. The first kit was an airplane, the *Heath Parasol*, offered in 1926. In 1931 Heath was killed in a test flight. After being run by Heath's widow for a few years, the bankrupt company was purchased by Howard E. Anthony in 1935. After the end of World War II he moved the company into the business of selling war surplus electronic parts and kits that made use of the readily available surplus parts. The first product was the O-1 oscilloscope kit in 1947 which sold for $39.50 and was a great success. It was followed by more test equipment including a VTVM, signal generator and condenser checker. Later the company expanded into amateur radio, hi-fi audio, and consumer products, all sold in kit form. Heathkit was renowned for the high quality of its assembly manuals and the premise of saving a significant amount of money by building electronics in kit form. While they may have been best known for their amateur radio products, the line of electronic test equipment was consistently successful throughout the kit era. Almost 500 different models of test equipment and accessories were offered between 1947 and 1992.

Howard Anthony was also killed in a plane crash in 1954. The company was purchased by Daystrom Inc, which owned other electronics companies including Weston, in 1958. In 1962 Heathkit was purchased by Schlumberger Limited, an oil exploration company. This period was the most successful one for Heathkit.

Wanting to expand into the new field of computers, the Zenith Radio Corporation purchased Heathkit in 1979. They focused more on computer products in the early 1980s and cut the budgets of the test equipment and amateur radio divisions. By the early 1990s kit sales were down and many of the kits were simply relabeled products designed by other companies. Looking for new markets, Heathkit even made a brief foray into furniture kits.

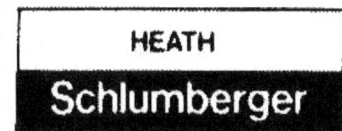

In 1989 Zenith, including Heathkit, was sold to Groupe Bull, a French-owned computer company. In 1992 Heathkit finally left the kit business entirely to focus on their educational products.

Groupe Bull sold Heathkit to an investor group called HIG in 1995, who then sold it to another investment group in 1998. During that time, the much smaller Heathkit continued as Heathkit Educational Systems. The rights to sell the legacy kit manuals and schematics were sold to an individual in 2008.

In 2011 Heathkit announced it was getting back into the kit business. It offered a couple of kits with more said to be planned. Despite generating a lot of hope in the kit building community, the offering was not particularly exciting or innovative: a car backup monitor and a pool alarm. This last gasp effort was not successful, and Heathkit finally declared bankruptcy in 2012.

Why did the once mighty Heathkit fail in the kit business? Analysts have suggested a number of factors. The early vacuum-tube based equipment was very labor intensive to assemble. Having the consumer assemble it enabled significant cost savings. However, with the trend toward solid-state and integrated circuits and eventually surface mount technology, most manufacturers moved to heavily automated assembly which meant there was little or no savings to be had by having the consumer assemble the product. The increased complexity of modern circuits also made kits more difficult to assemble and debug, increasing the chance of errors. Heathkit saw heavy competition from Japan and other Asian countries. The effort to develop the assembly procedures and write the renowned Heathkit manuals was significant, sometimes even more than the effort to design the product. By the 1990s the general public seemed to be less interested in electronics and building kits.

Ironically, in the last five years or so, there has been a resurgence in do-it-yourself technology including electronics. Now sometimes referred to as the "maker" movement, this has enabled a market for many Internet-based companies selling parts and kits. Companies like Elecraft are doing well (in a smaller niche market) selling amateur radio kits, and products like the low-cost Arduino and Raspberry Pi computing platforms are being used for innovative projects, including robotics (something Heathkit pioneered). But the market is very different from in the past, and I suspect Heathkit was just not up to the task of reinventing themselves in order to succeed in today's Internet-based world.

Is Heathkit gone for good? It would seem so, but over the years the company has changed ownership and had to reinvent itself several times. As I write this in 2013, the Heathkit web site contains a web page that says "The news of my death has been greatly exaggerated". Clicking on the page runs a detailed survey asking people about their interests and suggestions for possible Heathkit products, with a strong emphasis on amateur radio products as well as consumer kits. Details are scarce as the company is still in "stealth mode" but by the time you read this more information may have emerged.

What was it like to build a Heathkit? For most builders it was a very pleasant experience and a lot of fun. The renowned Heathkit manuals made it easy by breaking down the assembly into small steps where each step was checked off as it was completed. They made liberal use of pictorials to illustrate how all the parts were put together. The more complex kits often consisted of separate sub-assemblies like multiple circuit boards. The advent of printed circuit boards allowed building kits with larger part counts in less time and with less chance of error than with the older style of point-to-point wiring. This generally coincided with the introduction of solid-state components although some early tube-based kits like the V-7 VTVM introduced in 1955 used printed circuit boards. The manuals included a troubleshooting section that helped diagnose common problems. If you were still unable to get your kit working, you could contact Heathkit by phone, mail or in person at a store and get assistance from a technical consultant.

In addition to the kit-specific assembly manual, each kit came with the Heathkit kit builder's guide that covered advice on unpacking, identifying and sorting parts, and soldering techniques. Typically you only needed a soldering iron and solder and some basic hand tools like screwdrivers and side cutters to assemble a kit. Heath went to great lengths to minimize the need for any additional test equipment for adjustment and calibration. Some kits did require instruments, like a VTVM or oscilloscope, and if so this would be clearly indicated in the catalog. Some kits included their own calibration circuitry or test instruments.

Heath used high quality components and parts. Attractive cases and front panels gave the completed units a professional look. There was never a need to cut, sand, or paint. At the same time, they reduced overall cost by reusing the same or similar case designs for some series of units like test equipment and amateur radio products.

The complexity and difficulty of kits varied, starting on the low end with simple "one-evening kits". These were a good way to get introduced to electronics, learn to solder, and gain confidence before tackling a larger kit. They were ideal for young people or for a parent and child project.

The more challenging kits included products like color televisions which might involve more than ten circuit boards but were made easier with pre-assembled wiring harnesses and factory assembled and aligned tuner, power supplies, and the inclusion of a meter and dot generator for testing and alignment. Some of the amateur radio and test equipment was also quite complex to assemble, but the builders of these items tended to be more technical than those for the general consumer electronics kits.

Finally, the manuals for test equipment typically included extensive sections on operation and applications for the instrument after it was built. In the case of a dip meter or television alignment generator, this might be more than ten pages of text. Manuals also included a full schematic diagram and a theory of operation section explaining how the circuitry worked.

Early flyers and catalogs included full schematic diagrams for the products, although this became impractical when the product line became larger. Potential buyers could order just a manual before buying a kit to see the detailed product features and exactly what the assembly involved.

The most common cause of problems when building a kit was opens or shorts caused by poor quality soldering. Next most likely was errors in assembly, followed by one or more bad components. A fatal mistake in building a kit was to use acid-core solder or paste. Intended for plumbing use, the acid would soon eat away at the electronics and destroy it. Every Heathkit clearly noted this with a warning in the assembly manual.

Occasionally you will find a kit that has very bad soldering or where the builder may have abandoned construction, set it aside, and never got back to it. But based on kits seen on eBay and other sources, the vast majority of kits were successfully built and enjoyed by their proud builders for many years.

Illustration 3: 1957 Heathkit Catalog Page Featuring Actor Jackie Coogan And His Son

Chapter 2: An Overview of Heathkit's Test Equipment Lines

Over the years Heathkit used various product naming systems. Other than some early accessories, they followed a system that used one to three letters indicating a major product line, followed by a unique number. For example, the IM-18 denoted an instrument in the IM (meter) series. During the heyday they had a pretty consistent product naming scheme. Table 1 lists the major product series.

Some units, particularly test equipment, could be bought either in kit form or assembled. Some of the product lines indicate factory assembled or "wired" versions. For products that were offered assembled only, they were sometimes branded as "Heath", particularly in the later days when they were out of the kit business and the units were part of an educational product. Often the wired and kit versions used similar model numbers, such as SM-1212 and IM-1212, but this was not always the case.

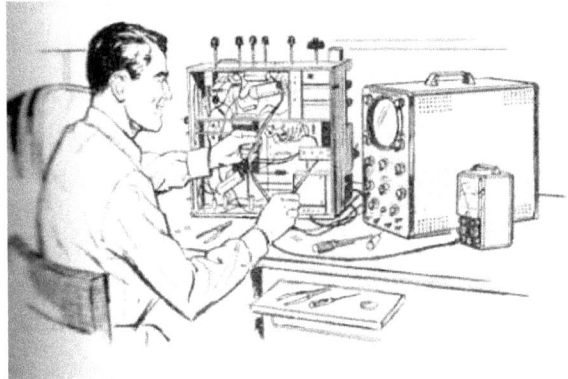

Heathkit offered a series of instruments branded as the Malmstadt-Enke Lab Components (EU/EUP/EUW series). Malmstadt and Enke were authors of the textbook *Digital Electronics for Scientists* and the instruments could be purchased along with the book. These were used by a number of universities that taught digital electronics to scientists. A similar partnership was a series of instruments marketed by Heathkit as *The Berkeley Physics Laboratory*.

Weston was the instruments division of Daystrom, the owner of Heathkit for a period of time, and some Weston meters were sold as factory assembled products under the Heath name. Weston was also a maker of tube testers, and at least one Weston tester was adapted to become a Heathkit model.

Heathkit offered some educational products that bundled instruments with a course, such as a VTVM. There were also some courses that had their own unique hardware, like the EK-1.

Most Heathkit products could be wired for 120 VAC 60 Hertz power for North America and 230 VAC 50 Hertz for European and international markets. A few products had specific versions just for the UK market.

Over the years, Heathkit test equipment went through a series of style changes, reflecting both the popular styles of the times and decisions by Heathkit's industrial designers. Sometimes the same equipment was reintroduced under a different model number that only differed in style or cabinet color. While it varied depending on the instrument, and there were many exceptions, I've noticed a few common style themes. Going from the oldest to newest equipment, ranging from the late 1940s to the early 1990s, I've observed the following styles generally used:

1. Battleship gray cabinet with red lettering. Often used black "chicken head" knobs.
2. Gray cabinet with white lettering. Gray plastic knobs.
3. Two-tone light and dark gray cabinet with white lettering. Black plastic knobs.
4. Beige cabinet with gray knobs having black centers.

5. Blue cabinet with white front panel. Black knobs, red for concentric knobs.

Early equipment tended to be taller than it was wide and incorporated a handle on top. The later style used wider cabinets with handles (if any) on the sides.

A note about measurement units, as this can be a possible area of confusion with older equipment. As an American company, Heathkit catalogs and manuals generally gave dimensions, weights and other measurements in what are now known as United States customary units (i.e. inches, pounds, degrees Fahrenheit). Later catalogs, particularly the Canadian versions, would often also list metric units (Canada officially adopted the metric system in the 1970s).

Over the years, the units used for various electrical properties have changed somewhat. The Hertz was officially adopted as the unit of frequency in 1960 but was not consistently used until the 1970s. Newer Heathkit manuals and catalogs would therefore use hertz (Hz), kilohertz (kHz), megahertz (MHz), and gigahertz (GHz). Prior to that, frequency was usually measured in the equivalent unit of cycles per second (cps), and sometimes just "cycles". Common units were kilocycles (KC) and megacycles (MC). The front panels of Heathkit instruments reflect the units that were common at the time. Very old radios sometimes display wavelength in meters rather than frequency, on their dial scales.

The units of resistance, capacitance, and inductance have been standardized as metric units from the beginning. However, for capacitance, the unit nanofarad (nF) for 10^{-9} Farad was not commonly used in North America until recently, and the picofarad (pF) for 10^{-12} Farad, was commonly referred to as the micromicrofarad (μμf) in some older manuals and catalogs. You may also see references to the word *condenser*. This is simply an older name for *capacitor*.

In this book I generally use the modern units, but when listing product specifications, they are reproduced exactly as in the original Heathkit manual or catalog entry, in some cases using the now obsolete measurement units.

Table 1: Major Heathkit Test Equipment Series

Series	Description
numbers	Adaptors and probes
AG	Audio Generators
AV	VTVMs
C	Condenser Checkers
ES	Power Supplies
ETI	Instruments from Educational Series
EU	Malmstadt-Enke Instruments
EUW	Malmstadt-Enke Instruments (wired)
G	Signal Generators
GD	Grid Dip Meters
IB	Impedance Bridges
IG	Signal Generators
IM	Meters
IN	Component Substitution Boxes
IO	Oscilloscopes
IP	Power Supplies
IR	Chart Recorders
IT	Miscellaneous Testers
O	Oscilloscopes
OM	Oscilloscopes
PK	Probes
PKW	Probes (wired)
PS	Power Supplies
S	Electronic Switches
SG	Signal Generators
SM	Factory Assembled Versions of IM Series
SO	Factory Assembled Versions of IO Series
SP	Factory Assembled Versions of IP Series
T	Signal Tracers
TC	Tube Checkers
TS	TV Sweep/Alignment Generators
TT	Tube Testers
V	VTVMs
VC	Oscilloscope Calibrators

You get highest quality at lowest cost
THE HEATHKIT® WAY!
Here are ten reasons why:

1. ✳ Building a Heathkit is easy—check-by-step instruction manuals make it virtually impossible for you to fail.

2. ✳ Building a Heathkit is quick—No complicated, technical jargon for you to decipher; at most, a Heathkit takes only a few evenings to assemble.

3. ✳ Building a Heathkit is economical—Mass production and purchasing economies are passed directly along to you, our customers.

4. ✳ Building a Heathkit is educational—As you build, you learn . . . more about electronics, more about your component units and when and where to add them.

5. ✳ Building a Heathkit is fun—Nothing quite equals the sense of achievement you receive when you successfully complete a Heathkit unit and "tune-in" for the first time.

6. ✳ Heath Company unconditionally guarantees that each Heathkit® product, whether assembled by our factory or assembled by the purchaser in accordance with our easy-to-understand instruction manual, must meet our published specifications for performance or your purchase price will be cheerfully refunded.

7. ✳ Your Heathkit is available on Convenient Credit—Our time payment plan makes it possible for you to order now . . . pay later. See order blank for complete details.

8. ✳ Your Heathkit is tops in quality—The very finest in electronic equipment comes to you in kit form from the Heath Company, for at Heath, quality is a matter of personal honor.

9. ✳ Heathkit Service is customer service—Our staff of technical experts is always ready to answer your questions or help you if you have any difficulty.

10. ✳ Factory Service by Skilled Technicians is always available for your Heathkit.

Chapter 3: Buying and Restoring Vintage Test Equipment

In this section I present some advice and tips on obtaining and restoring vintage test equipment. Most of this information is applicable to any vintage electronics, particularly if it is vacuum-tube based.

The traditional sources for finding old equipment are flea markets, amateur radio hamfests, and garage sales. Some of the best deals are often found here, but it is hit and miss. With the advent of the Internet much of this has moved to Kijiji, Craigslist, and of course eBay. On eBay it is possible to find almost anything, but prices tend to be higher, as you are essentially bidding against everyone else in the world who wants that particular item. You can find lots of advice on the Internet about shopping on eBay including techniques for bidding, sniping, etc. My suggestions are to examine the pictures and description of the item carefully and ask the seller lots of questions. Don't assume anything. Look at the seller's feedback. Set a limit for what you want to bid and try stick to it and don't let the excitement of bidding affect your judgment. Factor in the cost of shipping, insurance, customs and handling fees, and possible currency exchange. Be patient, and eventually you can usually get the item you want at a reasonable price, although it may take time. You can use the eBay saved searches feature to have the site e-mail you about items that match your criteria.

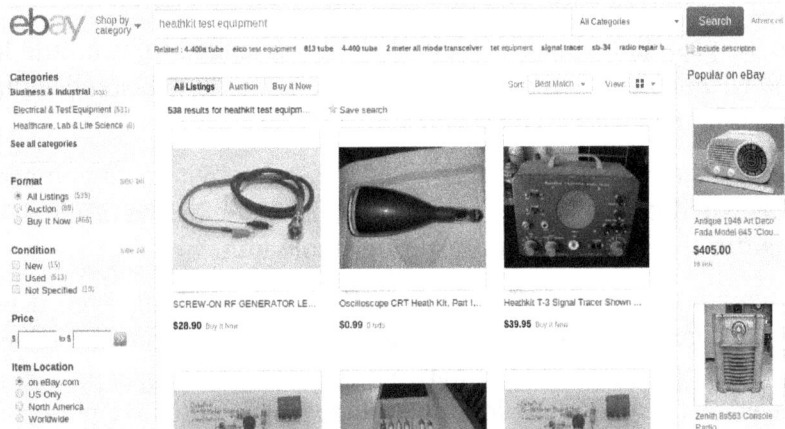

If you are up to the task of diagnosing and repairing old electronics, it is not particularly important if the unit is in working order. Often what a seller says is "working" means that when it was turned on a tube or pilot light came on. Look at the restoration and repair process as a challenge and source of entertainment – as one collector said when shown a unit that was not working: "I pay extra for that!". In fact, it is often best to ask sellers specifically *not* to power up old units, as this may cause damage to it.

Getting replacement parts such as resistors, capacitors, etc. is rarely a problem. Even vacuum tubes are readily available, although in my experience most problems are usually *not* due to a bad tube, unless it is missing, visibly broken, the wrong type, or has a burned out filament. Despite that, sellers will often say that a non-working unit "probably just needs a tube". What is harder to correct is missing unique mechanical parts like front panels, variable capacitors, and tuning dials. Beware of modifications, especially ones that cannot be easily reversed like holes in the chassis or entirely new circuit boards. Sometimes units are heavily corroded and may not be salvageable without a huge effort, like stripping it down completely to the bare chassis.

Electrolytic capacitors tend to fail after a few decades. The symptom in a radio is usually a loud hum. These can be replaced, although sometimes the old capacitors can be reformed using a capacitor tester or variable autotransformer. You may need to get these capacitors from a supplier that specializes in high-voltage replacement caps for vintage electronics. Paper capacitors, often sealed with wax, also tend to get electrically leaky over time and should be replaced. Many circuits placed bypass capacitors across the AC line or from line to ground. For equipment that you plan to operate on a regular basis, these should ideally be replaced with X1/Y2 rated capacitors that are designed to safely handle the surges and spikes that are encountered on the power line.

Older equipment was often not fused. You may want to add a suitable fuse in the AC line to protect the circuit from damage due to a fault such as a shorted transformer, filter capacitor, or rectifier tube.

Carbon resistors tend to change value over time and may no longer be within tolerance. They can also fail completely due to overheating. Ideally you should check all components and replace any that are bad.

Before powering up a unit, examine it first for any shorts, missing parts, etc. Then power it up slowly with a variable autotransformer, watching for any problems like smoke or arcing. Tube circuits have high voltage, some very high. Use caution when working on live circuits and use an isolation transformer to reduce the likelihood of a shock. A common safety rule is to put one hand in a pocket to reduce the chance of a fatal shock that would go through your body via the heart.

In the past, line voltages in North America tended to be a little lower than the 120 volts that is standard today. For this reason you may find that some measured voltages appear slightly higher than those listed in manuals and on schematic diagrams.

Unlike diagnosing a unit that failed in service, with a kit you can't necessarily assume it *ever* worked or was wired correctly. You may find errors made by the original kit builder decades ago. This all adds to the fun and satisfaction of restoring an old unit.

A common debate is whether to keep a unit original, even if not working, or to restore it to working order even if it means using some modern parts. Many units have had various modifications published over the years that may or may not improve performance. I recommend only to make mods that can be reversed (i.e. that don't involve drilling holes, etc.). My philosophy is to restore units to working order and to use them periodically and not just keep them on display on a shelf. Regular operation is generally good for the electronics and detects problems sooner. But for rare units, it might be worth keeping it as original as possible. Some restorers "restuff" capacitors by putting new parts inside the cases of the old ones, either metal cans for the electrolytics or paper reproductions of the cases of wax paper capacitors. I also know of people that have spent years tracking down parts from the original manufacturers and original date codes to keep a unit original – but this really only makes sense for an extremely rare unit.

The original Heathkit manuals often have critical information about operation and calibration. If you didn't get an original manual these can be obtained. Many partial manuals can be found for free on the Internet. You can also purchase full copies of the manuals from several sources – see Appendix A. The appendix also lists some web sites and mailing lists for vintage electronics including Heathkits. Often you can find useful information on specific models such as common problems and modifications.

I recommend you document what you find during restoration and what you do to the unit (if not for you, then maybe for the next person who works on it). Some of us are older and our equipment may outlive us. You may want to make an inventory of the more valuable items and let your family know what items shouldn't be left out on the curb on garbage day. Let them know what you wish to do with it (e.g. donate to ham club, etc).

Table 2 is a checklist of suggested steps to follow when restoring a piece of old equipment. Not all steps are mandatory but it is the process that I generally follow. I have found it useful to print off the checklist and refer to it during restoration.

The holy grail for some collectors is an unbuilt Heathkit still in the original box. These show up periodically on eBay, but expect to pay a king's ransom. Beware that it may be partially built and/or have missing parts. Buying a sealed box sight unseen could be risky. If you are successful in acquiring it, the question becomes whether to assemble it or keep it (possibly as an investment). If you do assemble an original kit, I encourage you to make a video of the process and put it on YouTube for others to appreciate. Oh, and if you have an unbuilt kit and aren't comfortable assembling it, I would be happy to do this for you ☺.

For those interested in collecting Heathkit test equipment, you might be wondering which models are the most collectible. It really depends – many collectors specialize in one type of equipment, such as VTVMs or tube testers, rather than test equipment in general. Many people also collect equipment of one type from different manufacturers, not just Heathkit. Some items are collectible more for nostalgia reasons, while others are desired because they continue to be useful as test equipment. Based on observations on eBay and other sources of used equipment like flea markets, hamfests, and Kijiji, I would divide the equipment into three categories of collectability. Note that this is for test equipment only, I'm not considering the other Heathkit product lines like amateur radio, audio, computers, or general consumer electronics products.

I would rate the following types of Heathkit test equipment in the category that has the highest desirability and attracts the highest prices:

1. Unbuilt kits, in general.
2. Older kits (i.e. prior to about 1960).
3. Oscilloscopes, either higher end ones or very old ones.
4. Tube testers.

Of these, an O-1 oscilloscope and V-1 VTVM would be very desirable as they were Heathkit's first products. I would include the following in the intermediate category of models that are desirable, but quite readily available at reasonable prices:

1. AF and RF signal generators.
2. Resistor and capacitor substitution boxes (with decade boxes being more desirable).
3. VTVMs.
4. Capacitor testers.
5. Frequency counters.
6. Dip meters.

Finally, items that would be the least desirable and collectible would, in my opinion, include the following:

1. High voltage power supplies.

2. Television servicing equipment.

3. Oscilloscope electronic switches, frequency scalers, and vectorscopes.

4. Scientific instruments.

5. Kits that have significant cosmetic issues (e.g. rust and corrosion, missing parts, major modifications).

6. Individual parts from kits (with a few exceptions for certain parts that often fail and/or are hard to obtain).

Original manuals are also reasonably valuable, at least compared to their original purchase price which was often on the order of one or two dollars. If you are looking for a specific manual and don't want a reproduction, you can likely find an original on eBay if you are prepared to wait patiently for a few months for one to show up.

Table 2: Restoration Checklist

1.	External visual inspection – make sure unit is complete with all components such as knobs, switches, etc. Look for any physical modifications such as additional switches or jacks.
2.	Take pictures of unit with digital camera including closeups of any complex assemblies. This will help if there are any issues when reassembling the unit.
3.	Internal visual inspection: check for missing, burnt, or damaged components. Look for any obvious circuit modifications.
4.	Obtain a copy of the manual, if missing. Note that due to design changes there may be differences between the unit being restored and the manual. Review the manual including the initial test, calibration, and operation sections.
5.	Check all components visually against values in the schematic and measure values of components when feasible (e.g. resistors and capacitors using a DMM). Note that values often can not be accurately measured in-circuit. Replace any parts which are faulty or significantly out of value. Replace wax paper capacitors if they are found to be leaky or suspect or if complete "shotgun" replacement is being done. Replace electrolytic capacitors that are leaky, out of value, or in any way suspect.
6.	Check tubes with a tube tester (if available) and note results. Verify that tube types match schematic and manual. Tubes which test weak may still be okay, but burnt out filaments or physically broken tubes will need to be replaced.
7.	Check line cord and replace if missing or unsafe. Optionally add fuse, if not present, to line circuit.
8.	Perform initial power up, using isolation transformer and variable autotransformer, if applicable. Watch for excessive current draw, smoke, or burning components. Check that tubes light up, pilot lights come on, and make initial check for proper operation of the unit.
9.	Remove and clean exterior of case and knobs using soap and water or stronger cleaner if necessary. Touch up or repaint case if desired and applicable.
10.	Clean interior chassis using paint brush and/or damp cloth. If cleaning tubes, use care to avoid removing marking. Areas with rust may require more extensive cleaning and disassembly.
11.	Clean switches and controls using contact cleaner. Lubricate any moving parts with oil or grease.
12.	Perform full power up test, checking all functions and making measurements as per manual, noting any problems or unusual behavior. Optionally check voltages against those found on the schematic. Use particular caution around any circuitry with high voltages or radio frequency power (e.g. transmitters).
13.	Diagnose and repair any problems found during testing.
14.	Perform calibration or alignment, if applicable, following the procedure in the manual.
15.	Perform final burn-in test, operating the unit for several hours to check for proper operation, any intermittent issues, overheating, or early component failures.
16.	Document the work performed, take pictures if desired, and enjoy your newly restored Heathkit.

Unusual Products: Here are some products offered by Heathkit that didn't fit into their usual product categories.

- GD-1150 Ultrasonic cleaner. Ideal for cleaning jewelry, watch parts, coins, dentures, etc.

- GD-1338 Ceiling fan and lamp. Sold as a kit, it could be assembled in less than an hour.

- WH-9121 Letter Quality Printer/Typewriter. An early daisy wheel printer that was also an electronic typewriter. Sold for $1,995.00 in 1982.

- GD-1026 Emergency Strobe Light. 12 volt operated strobe light that could be used as a highway emergency warning or signal light.

- IC-2006 Four Function Pocket Calculator. One of several products offered, it was a basic calculator in kit form. Sold for $29.95 in 1976.

- TD-1006 Color Organ. Popular in the 1980s disco era, when attached to a stereo system it flashed colored lights in unison with the music.

- PD-18 "Color Canoe". A metal tray for color film developing that used a rolling action to ensure consistent processing of film. Came in three sizes.

- GD-101 R/C Car. A 1/8 scale radio-controlled racing car kit featuring a gas engine and ran at scale speeds of up to 200 miles per hour. Cost was $49.95 in 1971. Engine and radio control equipment were not included but were available from Heathkit.

- GD-71 Telephone Amplifier. An early telephone hands-free conferencing unit in the days where telephones were rented and hard-wired to the wall. The handset rested in a cradle which amplified the audio through a speaker. Sold for $27.95 in 1961.

- GU-1810 Log Splitter. Powered by a five horsepower gas engine and utilizing a hydraulic pump, it could split logs up to 21" in length. Offered as a kit that could be built in one night, it sold for $589.95 in 1983.

- CA-1 CONELRAD Monitor. A chilling artifact of the 1950s cold war. CONELRAD was a system of alerting the public to a pending enemy attack (i.e. nuclear war). The CA-1 automatically turned off ham radio equipment that was plugged into it if a CONELRAD alert occurred, to avoid radio transmissions from being used by enemy missiles to home in on targets. It became obsolete in 1963 when CONRELRAD was replaced by the Emergency Broadcast System.

Chapter 4: Component Testers and Substitution Boxes

In this category are testers for components such as capacitors, diodes and transistors as well as component substitution boxes. Tube checkers and testers are covered in a later section of their own.

Substitution Boxes

Resistance and capacitance substitution boxes are not test instruments as such, but are useful for testing and experimenting. They allow selecting various values of resistance or capacitance, typically using a rotary switch. A given unit usually provides just resistance or just capacitance. Some units allow selecting any value over a given range using switches to select each significant digit of the value – these are known as *decade boxes*. Less expensive units may simply provide a number of representative values. An important specification is the tolerance or accuracy of the values. High accuracy units, such as 1%, are more expensive. The other key specification is the power rating. You need to ensure you do not exceed the power rating or the components will overheat and be damaged. Boxes with higher power ratings are more expensive. Substitution boxes are still manufactured, with low end units available for under $30 and high end decade boxes sometimes selling for over $1,000. When testing substitution boxes, the value of resistors may have drifted over the years. Also check for resistors that may have burned out when the power handling capability was exceeded. Capacitors are less likely to drift in value (although the original tolerance may have been quite high), but paper capacitors may have become electrically leaky over the years. Using a DMM with resistance and capacitance ranges it is easy to check for proper values. A capacitor checker (which can be an old Heathkit model) can test for leakage. Heathkit made both low-cost substitution boxes as well as decade resistance boxes. Table 3 lists the models of Heathkit substitution boxes. Some of the models were electrically identical and only varied in appearance and styling.

Table 3: Heathkit Resistor and Capacitor Substitution Boxes

Model	Description	First Year	Last Year	Comments
CS-1	Capacitor Substitution Box	1953	1962	18 values from .0001 to 0.22µF
DC-1	Capacitor Substitution Box	1951	1961	Decade box, 3 knobs
DR-1	Resistor Substitution Box	1956	1961	Decade box, 5 knobs
EU-30A	Resistor Substitution Box	1970	1970	Decade box, 7 knobs
EUW-28	Resistor Substitution Box	1967	1970	36 values from 15Ω to 10 MΩ
EUW-29	Capacitor Substitution Box	1967	1970	18 values from .0001 to 0.22µF
EUW-30	Resistor Substitution Box	1968		Same as IN-17
IN-11	Resistor Substitution Box	1961	1967	Decade box, 6 knobs
IN-12	Resistor Substitution Box	1962	1967	36 values from 15Ω to 10 MΩ
IN-17	Resistor Substitution Box	1967	1978	Decade box, 6 knobs
IN-21	Capacitor Substitution Box	1961	1967	Decade box, 3 knobs
IN-22	Capacitor Substitution Box	1962	1967	18 values from .0001µF to 0.22µF
IN-27	Capacitor Substitution Box	1967	1977	Decade box, 3 knobs
IN-37	Resistor Substitution Box	1967	1978	36 values from 15Ω to 10 MΩ

Model	Description	First Year	Last Year	Comments
IN-47	Capacitor Substitution Box	1967	1978	18 values from .0001µF to 0.22µF
IN-3117	Resistor Substitution Box	1991		Decade box, 6 knobs
IN-3127	Capacitor Substitution Box	1981		Decade box, 3 knobs
IN-3137	Resistor Substitution Box	1977	1981	36 values from 15Ω to 10 MΩ
IN-3147	Capacitor Substitution Box	1977	1981	18 values from .0001µF to 0.22µF
RD-1	Resistor Substitution Box	1950	1951	Decade box, 5 knobs
RS-1	Resistor Substitution Box	1951	1962	36 values from 15Ω to 10 MΩ

Capacitor Checkers

Capacitor (or condenser) checkers are a piece of test equipment specifically for testing capacitors. While functions varied among models, they could commonly measure the value of a capacitor, perform leakage tests, and sometimes measure power factor. Many units could also measure resistance and some could measure inductance using an external reference inductor. Early units would typically use a Wheatstone bridge circuit to measure resistance or capacitance and a "magic eye" tube as an indicator. The eye on the tube would open to indicate when the bridge was balanced, at which point the value could be read off of a dial. A typical Heathkit model was the IT-28. A lower end model, like the IT-22, could only perform tests for a capacitor being shorted or open. By the 1970s capacitor testers were less commonly used as more accurate LC impedance bridges, digital capacitance meters, and eventually DMMs with capacitance ranges become commonly available at low cost.

The leakage test is one feature of most capacitor testers that is not provided by modern test equipment, although it is less important than in the past as capacitors tend to be more reliable and used at lower voltages than in the days of vacuum tubes. Paper capacitors, often sealed with wax, can become electrically leaky after many years of use and the leakage test can detect this. Electrolytic capacitors rely on a chemical reaction to form an insulating layer. Old electrolytic capacitors that have not been used for a long time may need to be "re-formed" by slowly increasing the voltage applied to them. The leakage test mode is also useful for reforming these old capacitors and is a good reason to keep one of these units around for working on old equipment. Note that the voltages applied during the leakage test are somewhat dangerous: direct current at as much as 600 volts. You want to avoid touching the leads and should turn the voltage control down to zero after testing to discharge the capacitor. Otherwise it could be left charged at a high voltage for quite some time.

A common issue with these units is the eye tubes - they get dim over time. They are relatively expensive to replace because tubes that are still bright are now quite rare. Table 4 lists the models of capacitor testers offered by Heathkit.

Table 4: Heathkit Capacitor Checkers

Model	Description	First Year	Last Year	Comments
C-1	Capacitor Checker	1948	1950	Magic eye indicator
C-2	Capacitor Checker	1950	1951	Magic eye indicator
C-3	Capacitor Checker	1951	1960	Magic eye indicator
CT-1	Capacitor Checker	1957	1960	Magic eye, opens and shorts test only
IT-11	Capacitor Checker	1961	1988	Magic eye indicator

Model	Description	First Year	Last Year	Comments
IT-22	Capacitor Checker	1963	1967	Magic eye, opens and shorts test only
IT-28	Capacitor Checker	1968	1977	Magic eye indicator

Impedance Bridges

An impedance bridge is a higher quality version of the capacitor tester which typically can measure resistance, capacitance and inductance, and sometimes other parameters such as dissipation factor (D) and storage factor (Q). It often indicates bridge balance using a meter rather than an eye tube, and provides more accurate measurements than a cap tester. A typical Heathkit model was the IB-28. Table 5 lists the models of impedance bridges offered over the years. Most of these units were essentially the same design with only minor circuit and cosmetic differences.

Table 5: Heathkit Impedance Bridges

Model	Description	First Year	Last Year	Comments
IB-1	Impedance Bridge	1950	1950	Battery powered, wood case
IB-1B	Impedance Bridge	1951	1966	Battery powered, wood case
IB-2	Impedance Bridge	1951	1956	Measures R, L, C, D, Q. Built in 1 kHz generator
IB-2A	Impedance Bridge	1957	1967	Measures R, L, C, D, Q. Built in 1 kHz generator
IB-28	Impedance Bridge	1968	1976	Measures R, L, C, D, Q. Built in 1 kHz generator
IB-3128	Impedance Bridge	1977	1980	Same as IB-28 but different paint
IB-5281	Impedance Bridge	1977	1990	Measures R, L, C. Solid-state, battery powered.
IT-2240	Impedance Bridge	1989	1990	Digital benchtop

Transistor/Diode Testers and Curve Tracers

Transistor testers are something of a curiosity. When transistors first came out they were much more expensive and less reliable than today and perhaps it made economic sense to test them. Some old equipment may even have the transistors in sockets! Possibly it was simply a belief that since tubes needed testers, their solid state replacement, transistors, did as well. But while vacuum tubes wear out over time and lose gain and electron emission, transistors don't wear out as such. Today you will likely just replace a ten cent transistor if it is suspect, unless it is a rare type. Heathkit sold several models of transistor testers. The low end units like the IT-27 could check bipolar (NPN and PNP) transistors for shorts, leakage, opens, and DC current gain. They could also test semiconductor diodes. More expensive models like the IT-121 could also test other transistor types like FETs, SCRs, and triacs and more accurately measure DC gain. Curve tracers like the IT-1121 could display more transistor parameters as a graph on an oscilloscope. A particularly interesting product was the IT-2232 Component Tracer. It displayed a voltage-current plot of a component under test on a built-in CRT. Having two separate inputs, it could display the characteristics of a component under test overlaid against the curve of a known good component, allowing them to be compared. It was useful for both passive (i.e. resistor, capacitor, inductor) and active (transistor) components, in or out of circuit. I suspect it was inspired by a very similar commercial product called the Huntron Tracker, versions of which are still sold. Illustration 4 shows some testers from a 1971 catalog. Table 6 lists the models of testers offered by Heathkit.

Illustration 4: Three Transistor Testers from the 1971 Catalog

TRANSISTOR TESTERS

Heathkit Portable In-Circuit Transistor Tester . . .
Measures Transistor DC Beta In-Or-Out-Of-Circuit . . .
Measures Leakage Out-Of-Circuit . . .
Tests Diodes In-Or-Out-Of-Circuit . . .
All At A Market-Shattering Price of Only $24.95

• Compare to similar instruments selling for several times more • Tests transistors for DC gain in or out of circuit • Tests transistors out of circuit for Iceo and Icbo leakage • Tests diodes in circuit or out of circuit for opens and shorts • Identifies unknown diode leads • Identifies NPN or PNP type transistors • Matches NPN and PNP transistors • Cannot damage device or circuit even if connected incorrectly • Big 4½" 200 uA meter reads directly in DC Beta and Leakage • Two DC Beta ranges, 2-100 and 20-1000 • Expanded leakage current scale, 0-5000 uA, 1000 uA midscale • 10-turn calibrate control • Portable, battery powered • Long battery life from single "D" cell • Handy attached 3' test leads for in or out of circuit checks • Front panel socket for lower power devices • Rugged textured beige plastic case with attached positive locking cover, convenient handle, and lead storage compartment • Build it in less than two hours

PRICE BREAKTHROUGH FOR TRANSISTOR SERVICING. Heath engineers have done it again. This time they have taken the high cost of in-circuit transistor testers and brought them down to earth. The new Heathkit IT-18 has the facilities you need for fast, in-circuit transistor testing and it costs just a fraction of other brands.

VERSATILE TESTING. The new Heathkit IT-18 tests transistor DC Beta in-circuit or out-of-circuit from 2 to 1000 (this is the specification commonly used by manufacturers and equipment schematics to determine transistor gain). The IT-18 also will test in-circuit diode forward and reverse current, indicating opens, shorts, if not shunted by low impedance; tests diodes out-of-circuit if low impedance shunt is present. Emitter to collector (I_{ceo}) and base to collector (I_{cbo}) leakage current is measured out-of-circuit . . . range is 0 to 5,000 uA with 1000 uA mid-scale and 5 uA at first scale mark. Use the IT-18 to identify NPN or PNP devices, identify anode and cathode of unmarked diodes, match transistors of the same type or opposite types as used in complementary circuits.

EASY TO USE. You can depend upon the IT-18 to quickly locate defective devices or stages . . . you don't even need special device specifications, just connect the IT-18, adjust the ten-turn calibrate control and press-to-test transistor gain. No need to worry about damaging a circuit either, the IT-18 is safe to use even if connected incorrectly. Completely portable, the IT-18 is powered by a standard "D" cell . . . ready to serve you anywhere in the shop or in the field . . . and the meter is shorted for damping when not in use to prevent in-transit damage.

EASY TO BUILD, EASY TO BUY. You'll complete the IT-18 construction in less than two hours . . . and you'll congratulate yourself on getting the best buy ever in in-circuit transistor testers. Order your IT-18 now!

Kit IT-18, 4 lbs. $24.95

IT-18 SPECIFICATIONS—D. C. Beta: x1 range—2 to 100, x10 range—20 to 1000. Out-of-circuit accuracy: ±5%. In-circuit accuracy: Indicates good or bad (accuracy depends upon circuit being tested); Iceo (out-of-circuit only)—0-5000 uA; Icbo (out-of-circuitry only) 0-5000 uA diodes; forward or reverse current—0-5000 uA. Power: One standard "D" cell (not supplied). Dimensions: 8½" wide, 4¼" high, 7¼" deep (including handle). Net weight: 2¼ lbs.

Kit IT-18

$24.95

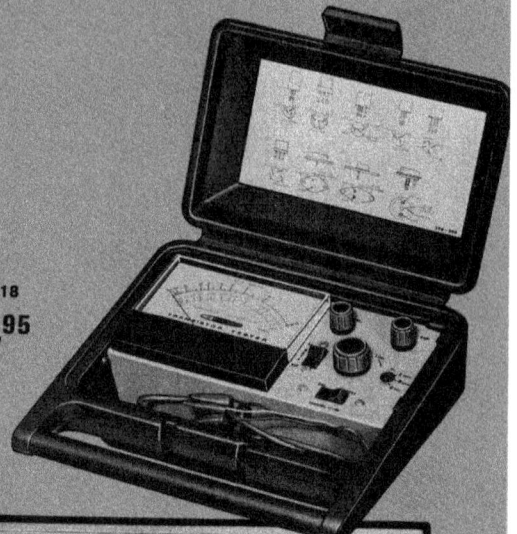

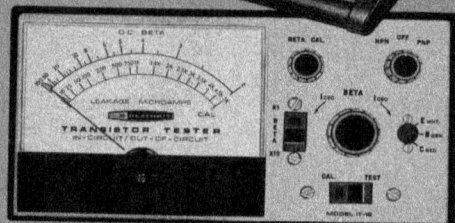

Big 4½" 200 uA meter with red and black scales . . . sensitive and easy to read.

Everything up front . . . all switches, calibrate control, and panel-mounted socket for easy out-of-circuit testing of small devices.

Improved Heathkit Laboratory Transistor Tester

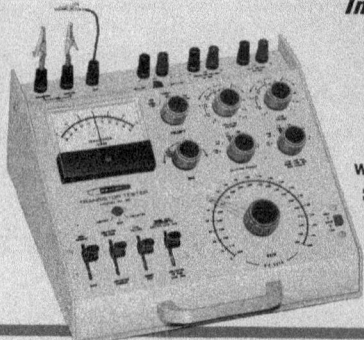

• Features the "new look" of Heathkit instruments • Measures current gain (Beta) up to 400, read on calibrated scales • Provides complete DC analysis of PNP & NPN types • 15 uA basic range for leakage tests (Icbo, Iceo) • Variable bias for setting collector current up to 15 amps • Internal power to 9 volts in 1.5 v. steps • Four lever switches for fast, easy test selection • Separate voltage & current range switches for both Gain & Leakage settings • Convenient carrying handle • Unmatched in performance, quality & price

Kit IM-36, 9 lbs., no money dn. $61.95
Assembled IMW-36, 11 lbs., no money dn. $92.50

IM-36 SPECIFICATIONS—Checks: Transistors up to 15 amps and diodes up to 1.5 amps. Tests: Shorts, DC Gain (Beta 0-200 & 200-400), Leakage (Icbo, Iceo), Diode Forward and Reverse Current. Meter: Current Ranges: 15 μA, 150 μA, 1.5 mA, 15 mA, 150 mA, 1.5 A, 15 A; Voltage Ranges: 1.5 V, 5 V, 15 V, 50 V, 150 V (100 K ohm/volt). Power supply: Internal seven 1.5 volt size D batteries provide 1.5, 3, 4.5, 6, 7.5, 9 volts collector supply for gain or leakage and 1.5 volts for bias; External, 0-50 volts for gain, 0-150 volts for leakage and 0-5 volts for bias. Bias control: Permits any collector current from 10 μA to 15 amps for gain tests. Gain control: 3% wire-wound control has calibrated scales to give DC Beta directly when meter is at null. Lever switch: Four spring return levers will individually select the following tests: Base Current, Gain, Collector Voltage, Collector Current, Leak Voltage, Short Test, Icbo and Iceo or diode currents. Two voltage selector switches: Gain and Leak Test Voltage can be individually preset at different voltage levels. Two current range switches: Collector and Leak currents can be individually preset on different adjacent ranges. Transistor and diode connections: Universal transistor socket and binding posts. External power supply connections: Binding Posts for Gain, Leak and Bias supplies.

Kit IM-36

$61.95

Wired IMW-36

$92.50

Heathkit Transistor/Diode Checker—Time Saving, Handy & Portable

• Features the new "look" in Heathkit instrumentation • Satisfies most servicing needs • Compact—handy for test bench or service kit • Checks both high and low power transistors • Checks PNP and NPN types • Checks for shorts, leakage, open element, and current gain • Checks forward and reverse current on diodes • Built-in switching eliminates transferring test leads • Completely portable—operates on two standard "C" flashlight cells • Plug-in leads with alligator clips for checking transistors which do not fit socket • Easy to use

Kit IT-27, 2 lbs. $6.95

IT-27 SPECIFICATIONS—Transistor test: Leakage, short, open, and current gain. Diode test: Forward and reverse current. Also serves as a continuity checker. Switches: Forward-Reverse/PNP-NPN, Diode/Hi-Lo, Leakage-Gain. Power supply: Self-contained, two 1.5 volt size "C" flashlight cells (not supplied). Dimensions: 3⅛" H x 3½" W x 3¾" D.

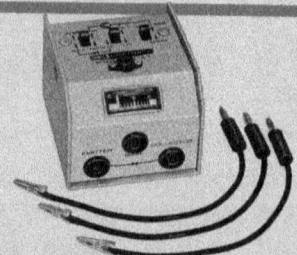

Kit IT-27

$6.95

Kits On These 2 Pages Are Mailable Except TT-1A Shipped REA Or Motor Freight

47

Table 6: Heathkit Transistor/Diode Testers and Curve Tracers

Model	Description	First Year	Last Year	Comments
IM-30	Transistor Tester	1961	1967	Tests bipolar transistors for shorts, gain, leakage
IM-36	Transistor Tester	1968	1974	Tests bipolar transistors for shorts, gain, leakage
IT-10	Transistor Checker	1961	1967	Tests bipolar transistors for shorts, gain, leakage
IT-18	Transistor Tester	1968	1978	Tests bipolar transistors for shorts, gain, leakage
IT-27	Transistor Checker	1967	1978	Tests bipolar transistors for shorts, gain, leakage
IT-121	Transistor Tester	1973	1976	Tests bipolar, FETs, SCRs, triacs, etc. Gain and transconductance.
IT-1121	Curve Tracer	1974	1977	Tests transistors, diodes. Used with scope.
IT-2127	Transistor Tester	1978	1981	Same as IT-37
IT-2232	Component Tracer	1984	1990	Built in CRT
IT-3118	Transistor Tester	1977	1979	Same as IT-18
IT-3120	Transistor Tester	1977	1989	Tests bipolar, FETs, SCRs, triacs, etc. Gain and transconductance.
IT-3121	Curve Tracer	1978	1983	Tests transistors, diodes. Used with scope.
IT-3127	Transistor/Diode Checker	1978	1983	Simple diode and transistor tester

IC Testers

The IT-7400 (Illustration 5) was Heathkit's only IC (Integrated Circuit) tester. It is something of a curiosity – it can test digital logic gates by applying logic levels to input pins and displaying output levels. It supported several different families of digital gates, and can also test 7-segment LEDs. In practice it must have been rather tedious to use as it would require many different switch settings to test even a simple logic gate. The testing it could perform would be limited to simple logic gates and flip-flops and only at DC levels. I suspect it was of more use as a learning tool, and Heathkit specifically mentioned that it was ideal for use with their Digital Electronics Course. Personally, I would like to own one if only because of all the switches and lights!

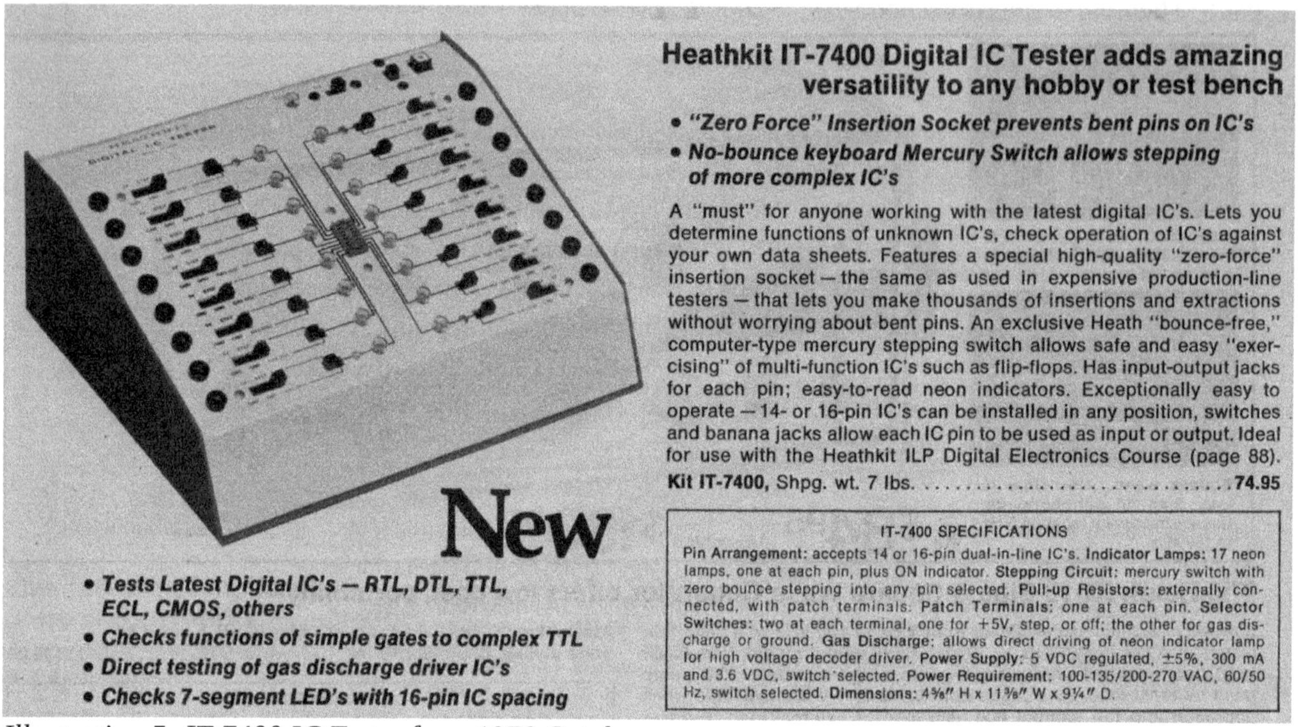

Heathkit IT-7400 Digital IC Tester adds amazing versatility to any hobby or test bench

- *"Zero Force" Insertion Socket prevents bent pins on IC's*
- *No-bounce keyboard Mercury Switch allows stepping of more complex IC's*

A "must" for anyone working with the latest digital IC's. Lets you determine functions of unknown IC's, check operation of IC's against your own data sheets. Features a special high-quality "zero-force" insertion socket — the same as used in expensive production-line testers — that lets you make thousands of insertions and extractions without worrying about bent pins. An exclusive Heath "bounce-free," computer-type mercury stepping switch allows safe and easy "exercising" of multi-function IC's such as flip-flops. Has input-output jacks for each pin; easy-to-read neon indicators. Exceptionally easy to operate — 14- or 16-pin IC's can be installed in any position, switches and banana jacks allow each IC pin to be used as input or output. Ideal for use with the Heathkit ILP Digital Electronics Course (page 88).
Kit IT-7400, Shpg. wt. 7 lbs. **74.95**

IT-7400 SPECIFICATIONS
Pin Arrangement: accepts 14 or 16-pin dual-in-line IC's. Indicator Lamps: 17 neon lamps, one at each pin, plus ON indicator. Stepping Circuit: mercury switch with zero bounce stepping into any pin selected. Pull-up Resistors: externally connected, with patch terminals. Patch Terminals: one at each pin. Selector Switches: two at each terminal, one for +5V, step, or off; the other for gas discharge or ground. Gas Discharge: allows direct driving of neon indicator lamp for high voltage decoder driver. Power Supply: 5 VDC regulated, ±5%, 300 mA and 3.6 VDC, switch selected. Power Requirement: 100-135/200-270 VAC, 60/50 Hz, switch selected. Dimensions: 4⅜" H x 11⅜" W x 9¼" D.

New

- *Tests Latest Digital IC's — RTL, DTL, TTL, ECL, CMOS, others*
- *Checks functions of simple gates to complex TTL*
- *Direct testing of gas discharge driver IC's*
- *Checks 7-segment LED's with 16-pin IC spacing*

Illustration 5: IT-7400 IC Tester from 1976 Catalog

In Depth: The IT-11 Capacitor Checker

In this section we'll take a closer look at one particular model, the Heathkit IT-11 capacitor checker. Illustration 6 shows the front panel. The unit measures resistance, capacitance, and capacitor leakage. It can also measure inductance with an external reference and transformer turns ratio. It uses a "magic eye" tube indicator rather than a meter.

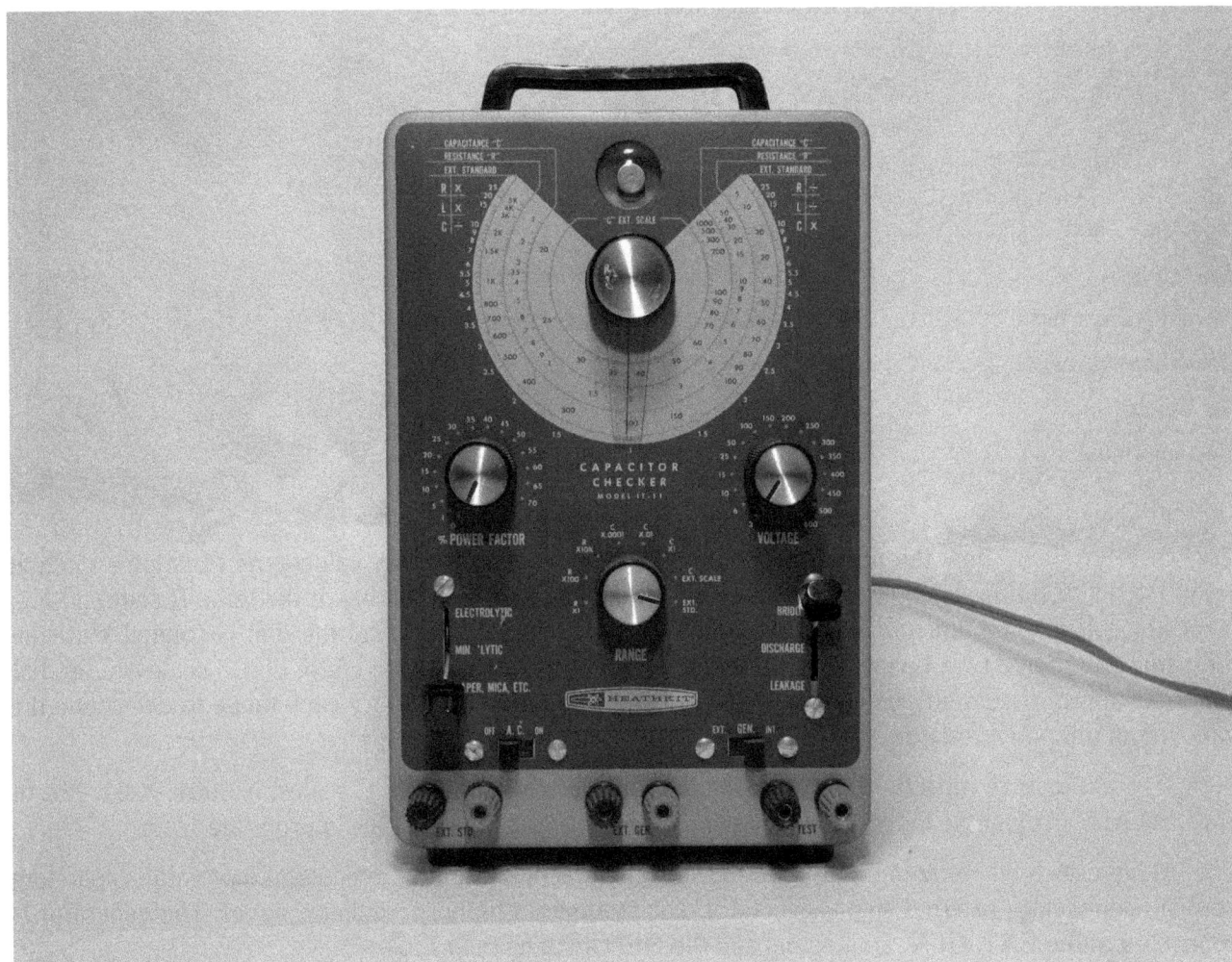

Illustration 6: IT-11 Front View

It was sold as a kit from 1961 to 1968, when it was replaced by the IT-28 which was almost identical except for a different color scheme. The IT-28 was sold until 1977. This unit sold new for between 30 and 40 dollars in the 1960s. The specifications are listed below:

SPECIFICATIONS:

Test Circuit . . . . AC bridge, powered by an internal 60 cycle supply or by an external audio generator with 10 volts output. Upper frequency limit: 10 kc.

Capacitance, 5 Ranges . . . . 10 μμf to 0.005 ufd.

.001 μfd to .5 μfd.

.1 μfd to 50 μfd.

20 μfd to 1000 μfd.

External standard (comparison bridge; maximum ratio 25:1)

Capacitor Leakage . . . . DC test voltages from 3 to 600 volts in 16 steps

Resistance, 4 Ranges . . . . 5 Ω to 5000 Ω.

500 Ω to 500 K Ω.

50 K Ω to 50 megohms.

External standard (comparison bridge, maximum ratio 25:1)

Inductance Check . . . . External standard only.

Power Supply . . . . Transformer-operated, half-wave rectifier.

Power Requirements . . . . .105-125 volts AC, 50/60 cycles, 30 watts.

Dimensions . . . . .9-5/8" high x 6-5/8" wide x 5" deep.

Net Weight . . . . 5 lbs.

Shipping Weight . . . . 7 lbs.

Let's briefly review how the unit was operated. To measure component values, the function switch is set to BRIDGE mode. To measure resistance, the range switch is set to one of the three R settings: X1, X100, or X10K. The unit under test is connected to the TEST jacks and the dial is rotated until the indicator on the eye tube opens. The value is then read off of the RESISTANCE "R" scale on the dial using the appropriate multiplier for the scale selected. You may need to try each range to determine the value of an unknown resistor.

Similarly, to measure capacitance you set the RANGE to one of the four C scales: X.0001, X.01, X1, or EXT. SCALE. Adjust the dial until the eye opens and read the value off the appropriate scale.

The bridge uses a 60 Hertz signal for testing. For more accurate measurements of low value capacitors you can connect an external signal generator that produces a higher frequency signal. The generator is connected to the EXT. GEN. connectors and the GEN switch set to EXT.

Electrolytic capacitors can also be checked for power factor. By turning the POWER FACTOR knob until the eye opens to a maximum amount you can read the power factor off the knob. Power factor is a measure of the equivalent series resistance (ESR) of a capacitor, and lower values are generally better.

To test capacitors for leakage the mode switch is set to LEAKAGE. The type of capacitor should be selected using the type switch, one of ELECTROLYTIC, MIN. 'LTIC (for miniature electrolytic capacitors), or PAPER, MICA, ETC. Set VOLTAGE to minimum and connect the capacitor under test. Now increase the voltage a step at a time. If the capacitor is not leaky the eye should remain open up to the rated voltage of the capacitor. Old wax paper capacitors will often show as very leaky. Note that this applies up to 600 volts DC across the device under test. While low current, it can produce an annoying shock if you touch the terminals!

After testing for leakage it is good practice to set the function switch to the DISCHARGE position before removing it. This will ensure the capacitor is not left charged to a high voltage.

The unit can also measure a value against an external reference component. While this works for resistance and capacitance, it is particularly useful for measuring inductance using an external inductor as a reference. The reference device is connected to the EXT. STD. terminals and the RANGE switch set to EXT. STD. When the bridge is balanced (eye open), the dial indicates the ratio of the value of the component under test versus the external standard component. It can also be used to determine the turns ratio between two windings of a transformer.

Illustration 7 shows the inside of the unit. It is quite a simple design using three tubes, a 6AX4 rectifier in the power supply, a 6BN8 triode/dual diode, and a 6E5 eye tube.

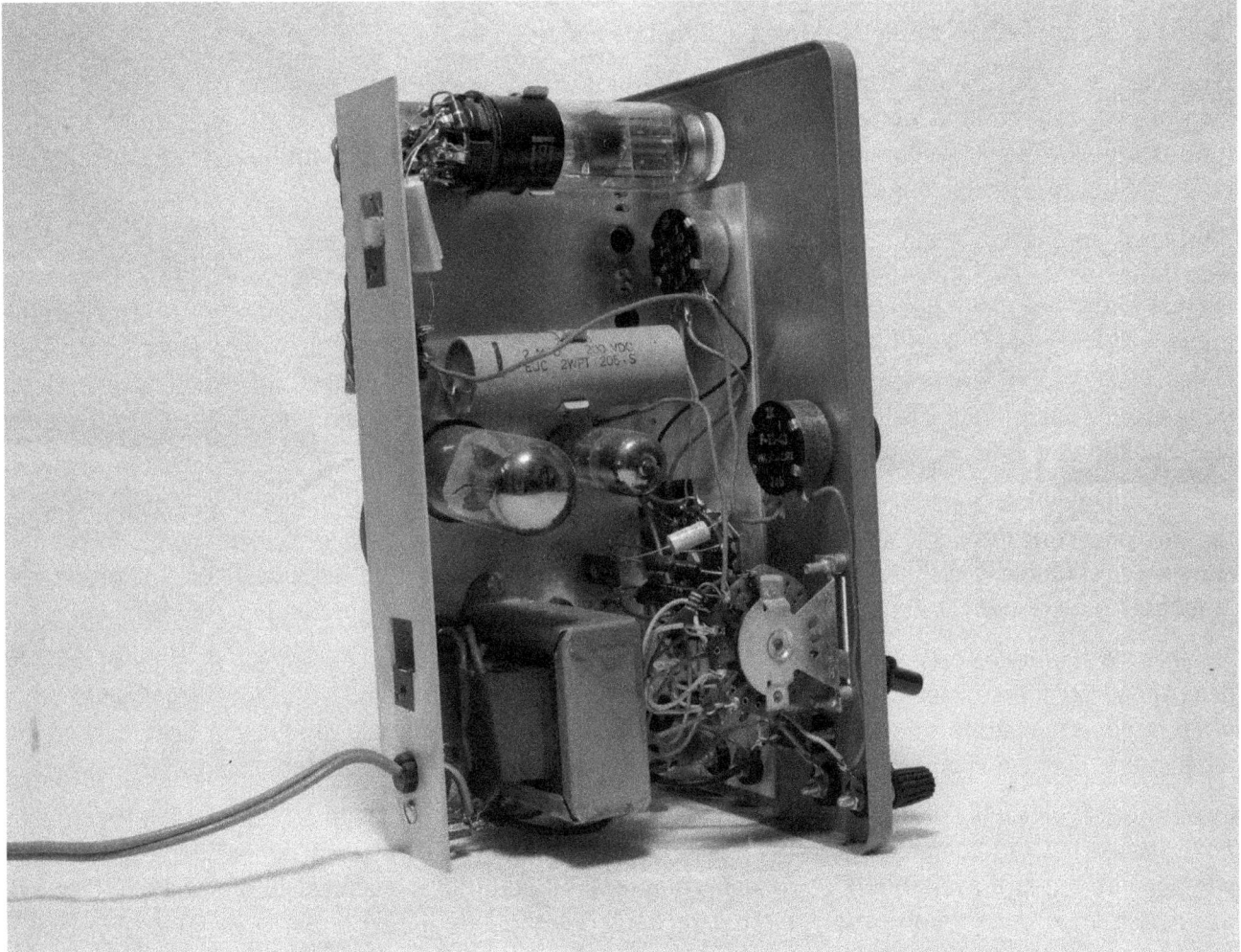

Illustration 7: IT-11 Inside View

I bought this unit used on eBay and it came with the original manual. It was working and only needed minor cleaning. The eye tube is still quite bright. Overall, the instrument is not very accurate but was useful for hobbyists and repairmen. It was much less expensive than laboratory instruments of the time. It is still useful for leakage testing, reforming electrolytic capacitors, and just plain nostalgia. This type of instrument has today been mostly replaced by much more accurate multimeters and LCR meters.

Chapter 5: Frequency Counters

A frequency counter is an electronic test instrument used for measuring frequency, period (the inverse of frequency), or counting the number of events, such as pulses. There is a need for accurate frequency measurement in many areas of electronics, most notably radio circuits. While frequency can be measured using an oscilloscope or grid dip meter (both covered in later chapters), they are limited in accuracy and in the upper range of frequency that can be measured.

Frequency is a physical quantity that can be measured with very high accuracy. While a meter that can measure voltage, current, or resistance with an accuracy of 1% or better is considered quite good, a typical digital frequency counter is rated with accuracy in the parts per million, with a resolution ranging from five to eight significant digits.

The term frequency counter generally refers to a *digital* frequency counter that uses counting techniques to measure frequency and indicate it numerically on a display. Early digital frequency counters used vacuum tubes, sometimes using specialized decade counter tubes, but were limited in the upper frequency that could be measured. The advent of digital integrated circuits, most notably TTL chips, made it practical to implement digital frequency counters that support measurement into radio frequencies. Early units used neon Nixie tubes for a display, later switching to light-emitting diodes (LEDs), and then liquid crystal displays (LCDs).

Most frequency counters operate by counting the number of transitions in the input signal over a specific period of time, known as the *gate time*. This can then be converted to a frequency value and shown on a numeric display. Measurements are referenced to an internal oscillator known as the *timebase*, and accuracy is dependent on the timebase being accurately calibrated.

Heathkit typically offered several models of counters at various price ranges. Price was mostly determined by the counter's upper frequency limit, precision (digits) and accuracy. Higher end units often used temperature controlled crystal ovens to minimize timebase frequency drift caused by temperature changes. Some provided additional features like period measurement and multiple inputs.

One of the few Heathkits that I actually built new from a kit was for my high school electronics class in 1976. Restoring old counters is usually quite straightforward -- all of them are solid-state and most components are still available. The exceptions would be getting replacement Nixie or LED displays that match the originals. Fortunately, these parts don't often fail.

In order to be accurate it is important to calibrate them – this can be done simply and very accurately using a shortwave receiver tuned to a frequency standard radio station such as WWV. Some early models had an optional external frequency scaler accessory that would extend the frequency range of the counter. Later models would often include an internal prescaler circuit. Illustration 8 shows a couple of frequency counter models from 1982. Table 7 lists the various models offered by Heathkit.

34

Test and align anywhere with the IM-2400 Counter

$279⁹⁵

- Battery operated for in-field testing
- Rechargeable batteries for low-cost operation
- Two frequency ranges and time bases

Now you can accurately test and align mobile radio equipment in cars, trucks, aircraft, boats — anywhere you want, with the portable IM-2400 Digital Frequency Counter.

It features a **10 MHz crystal-oscillator** that insures stable and accurate frequency monitoring through both 50 Hz-50 MHz and 40 MHz-512 MHz ranges. With a typical sensitivity of 10 millivolts, RMS, and a guaranteed sensitivity of 25 millivolts, RMS, the IM-2400 lets you measure all types of signals — including weak ones. And you can read those frequencies easily on the LED display with automatic decimal placing and zero blanking.

The IM-2400 is completely portable. It has rechargeable, nickel-cadmium batteries for hours of readings on one charge. The batteries are located inside the housing to eliminate awkward external battery packs. For bench operation, you can get one of the optional Battery Eliminator/Chargers (next column) to oper-

ate the IM-2400 directly off line current.
Kit IM-2400, Shpg. wt. 1.5 Kg 279.95

Swiveling Telescopic Antenna for the IM-2400. Also can be used on 2-meter transceivers. This chrome-plated brass antenna, with its right-angle and telescoping capability, gives you improved performance and better sensitivity. Includes BNC connector.
SMA-2400-1, Assembled, Shpg. wt. 1 Kg 16.95
120 VAC Battery Eliminator/Charger for IM-2400.
PS-2404, Shpg. wt. 1 kg 11.95

IM-2400 SPECIFICATIONS: **Input Impedance:** 50 Hz to 50 MHz range — 1 megohm shunted by less than 20 pF; 40 MHz to 512 MHz range — 50 ohms. **Input Protection:** 50 Hz to 50 MHz range — 150 volts RMS to 100 kHz, derating to 10 volts RMS at 50 MHz; 40 MHz to 512 MHz range — 5 volts RMS. **Setability:** +1 ppm. **Temperature Stability:** +10 parts ppm from 0°C to 40°C. **Gate Time:** 1.0 or 0.1 seconds, switchable. **Resolution:** 50 Hz to 50 MHz range — ±10 Hz at time base set at 1.0 second, ±100 Hz at time base set at 0.1 second; 40 MHz to 512 MHz range — ±100 Hz at time base set at 1.0 second, 1 kHz at time base set at 0.1 seconds. **Power Requirement:** Five 1.5 VDC, rechargeable nickel cadmium cells (included), or 120 VAC with optional PS-2404, or 240 VAC with optional PS-2405. **Dimensions:** 1.63″ H x 3.38″ W x 8.38″ D.

Here's a low-cost Digital Frequency Counter for your bench

$249⁹⁵

- Switchable ranges for accurate, high-resolution readouts
- Excellent typical 10 mV sensitivity
- Big, easy-to-read, 8-digit LED display with automatic decimal placement

If you're looking for an inexpensive, but highly accurate frequency counter for your bench, then check out this IM-2410 Digital Frequency Counter.

Accuracy and stability unmatched for its low-price. The IM-2410 features two frequency ranges — from 10 Hz to 50 MHz, and from 20 MHz to 225 MHz — for increased accuracy and better resolution. A two-position time gate lets you choose either 0.1 sec or 1.0 seconds for even better resolution. And the crystal-controlled time base gives you good long-term stability and accuracy you would expect to find on more expensive counters: ±1 ppm.

Front-panel operation for easy bench use. The IM-2410 has one BNC input for fast, direct frequency countings. For non-direct counts, you can attach the optional SMA-2400-1 Swivel Antenna (described above). In addition, all the IM-2410's operation switches are on the front panel for easy reach. The 8-digit display is easy-to-read, and shows all frequencies in MHz for higher resolution (as fine as 1 Hz at 50 MHz, 10 Hz at 225 MHz). A cabinet stand props the IM-2410 so that the display is positioned at a comfortable viewing angle.

Interference-free cabinet construction. The IM-2410's housing is made of metal for greater durability and better RFI (Radio Frequency Interference) shielding. All in all, the IM-2410's money-saving price, accuracy, and simplified operation make it a valuable instrument buy for engineer, technician, and hobbyist alike. Dimensions: 3.38″ H x 7.25″ W x 9.5″ D.

Kit IM-2410, Shpg. wt. 3 Kg 249.95

IM-2410 SPECS: **Max. Sensitivity:** 25 mV, RMS. **Input Impedance:** 1 MΩ shunted by less than 24 pF. **Input Protection:** 150 VAC, up to 100 kHz — derating to 5 VAC from 160 MHz to 225 MHz. **Time Base Frequency:** 3.58 MHz. **Temperature Stability:** +10 ppm, from 0° to 40°C. **Power Req:** 120/240 VAC, 50/60 Hz.

FM Deviation Meter helps check transmitter performance

$279⁹⁵

- Measures both FM Transmitters and signal generators
- Two front jacks for fast oscilloscope hookup

The IM-4180 lets you check transmitter and signal generator performance by measuring the peak FM deviation between 25 and 1000 megahertz. A BNC jack allows external and direct-line monitoring. An 8-ohm jack allows you to listen to deviation through a speaker or a pair of headphones.

All the jacks and controls of the IM-4180 are easy-accessible from the front panel. The controls include: four push button switches for selecting deviation modulation ranges from 2 to 75 Hz; course and fine- tuning controls for locking in difficult UHF FM

signals; a level control to adjust the meter sensitivity, and an audio output adjustment control.

The IM-4180 features a blue and white, 5″ H x 10.31″ W x 7.19″ D cabinet, and runs on 10 AA batteries (carbon-zinc, alkaline or nickel-cadmium — not included) for remote-location monitoring. The optional 120/240 VAC IMA-4180-1 Battery/Eliminator lets you use the IM-4180 without draining the batteries.
Kit IM-4180, Shpg. wt. 3.5 Kg 279.95

Battery Charger/Eliminator for the IM-4180.
Kit IMA-4180-1, Shpg. wt. 1 Kg 54.95

Swiveling Telescoping Antenna for use with IM-4180. For monitoring transmitter and generator performance without connecting IM-4180 to equipment.
SMA-2400-1, Assembled, Shpg. wt. 1 Kg 16.95

IM-4180 SPECS. **Carrier Freq. Ranges:** Fundamental, 25 to 50 MHz; **Deviation Ranges:** 0-2, 0-7.5, 0-20, & 0-75 kHz. **Accuracy:** ±3%, full scale. **Input Imped.:** 50Ω. **Max. Input:** 5 V. **Scope Output:** 13 mV/kHz peak.

Illustration 8: 1982 Catalog Page Featuring Two Frequency Counters

Table 7: Heathkit Frequency Counters

Model	Description	First Year	Last Year	Comments
ETI-7040	Frequency Counter	1990	1990	8 digit, factory assembled, Heath Zenith
IB-101	Frequency Counter	1970	1972	5 digit, 15MHz, Nixie display
IB-102	Frequency Scaler	1971	1975	Extends range of IB-101
IB-1100	Frequency Counter	1973	1975	5 digit, 30MHz, Nixie display
IB-1101	Frequency Counter	1972	1975	5 digit, 100MHz, Nixie display
IB-1102	Frequency Counter	1972	1977	8 digit, 120 MHz, Nixie display
IB-1103	Frequency Counter	1973	1977	8 digit, 180 MHz, Nixie display
IM-2400	Frequency Counter	1980	1991	7 digit, 512 MHz, hand-held
IM-2410	Frequency Counter	1980	1992	8 digit, 225 MHz
IM-2420	Frequency Counter	1980	1990	8 digit, 512 MHz, two inputs
IM-4100	Frequency Counter	1975	1979	5 digit, 30 MHz, period and totalize
IM-4110	Frequency Counter	1977	1979	8 digit, 110 MHz
IM-4120	Frequency Counter	1977	1979	8 digit, 250 MHz
IM-4130	Frequency Counter	1977	1979	8 digit, 1 GHz
SM-104	Frequency Counter			No details known
SM-105	Frequency Counter	1971	1971	80 MHz
SM-105A	Frequency Counter			80 MHz
SM-2372	DMM/Frequency Meter	1989	1992	20MHz, assembled
SM-2410	Frequency Counter			Assembled version of IM-2410
SM-2420	Frequency Counter			Assembled version of IM-2420
SM-2440	Frequency Counter	1989	1990	Assembled version of IM-2440
SM-4100	Frequency Counter			Assembled version of IM-4100
SM-4190	Frequency Counter			Assembled version of IM-4190

In-Depth: The IM-2410 and IM-2420 Frequency Counters

Let's take a detailed look at two representative Heathkit digital frequency counters, the IM-2410 and IM-2420.

IM-2410 Frequency Counter

The IM-2410 represented Heathkit's lower end digital frequency counter and was sold from 1980 through 1992. I've seen it listed as having a retail price of $119.95. My 1982 Canadian Heathkit catalog lists it at a price of $249.95. Illustration 9 shows the front panel of the unit.

Illustration 9: IM-2410 Front View

It can measure frequency up to 225 MHz. There is a single input using a BNC connector. A switch selects between 10 Hz to 50 MHz and 20 MHz to 225 MHz frequency ranges. Another switch selects between 0.1 second and 1 second gate times. Full specifications are listed below:

IM-2410 SPECS: Max. Sensitivity: 25 mV, RMS. **Input Impedance:** 1 MΩ shunted by less than 24 pF. **Input Protection:** 150 VAC, up to 100 kHz – derating to 5 VAC from 160 MHz to 225 MHz. **Time Base Frequency:** 3.58MHz. **Temperature Stability:** +10 ppm, from 0° to 40°C. **Power Req:** 120/240 VAC, 50/60 Hz.

The display shows eight significant digits on red 7-segment LEDs. The decimal point moves automatically to indicate frequency in MHz. It has a metal case with rubber feet and a tilt-up stand. It runs on AC power only. There is an adjustment on the back to calibrate the frequency of the internal oscillator.

Illustration 10: IM-2410 Inside View

Illustration 10 shows a view of the inside of the unit. Most circuitry is on a single large printed circuit board with a second smaller PCB for the display. It is all solid state using mostly 7400 series integrated circuits. The internal clock oscillator uses a 3.579545 TV color burst crystal. It is all discrete logic; no microprocessor is used.

I bought my unit on eBay in 2009. I needed a frequency counter for aligning tube radios and wanted to get a Heathkit as they are reliable and easy to repair. This unit goes up to 225 MHz, which is more than adequate for just about any radio work I do. It worked when received and all I did was calibrate it against a signal generator that was referenced to the WWV frequency standard station. I keep it on my test bench and use it regularly.

IM-2420 Frequency Counter

The IM-2420 was a higher end counter with more features. It was sold as a kit and was offered from 1980 through 1990. My 1982 Canadian catalog lists it at $449.95. It measures frequency up to 512 MHz. It can also measure period, which is useful for very low frequencies. Illustration 11 shows the front of the unit.

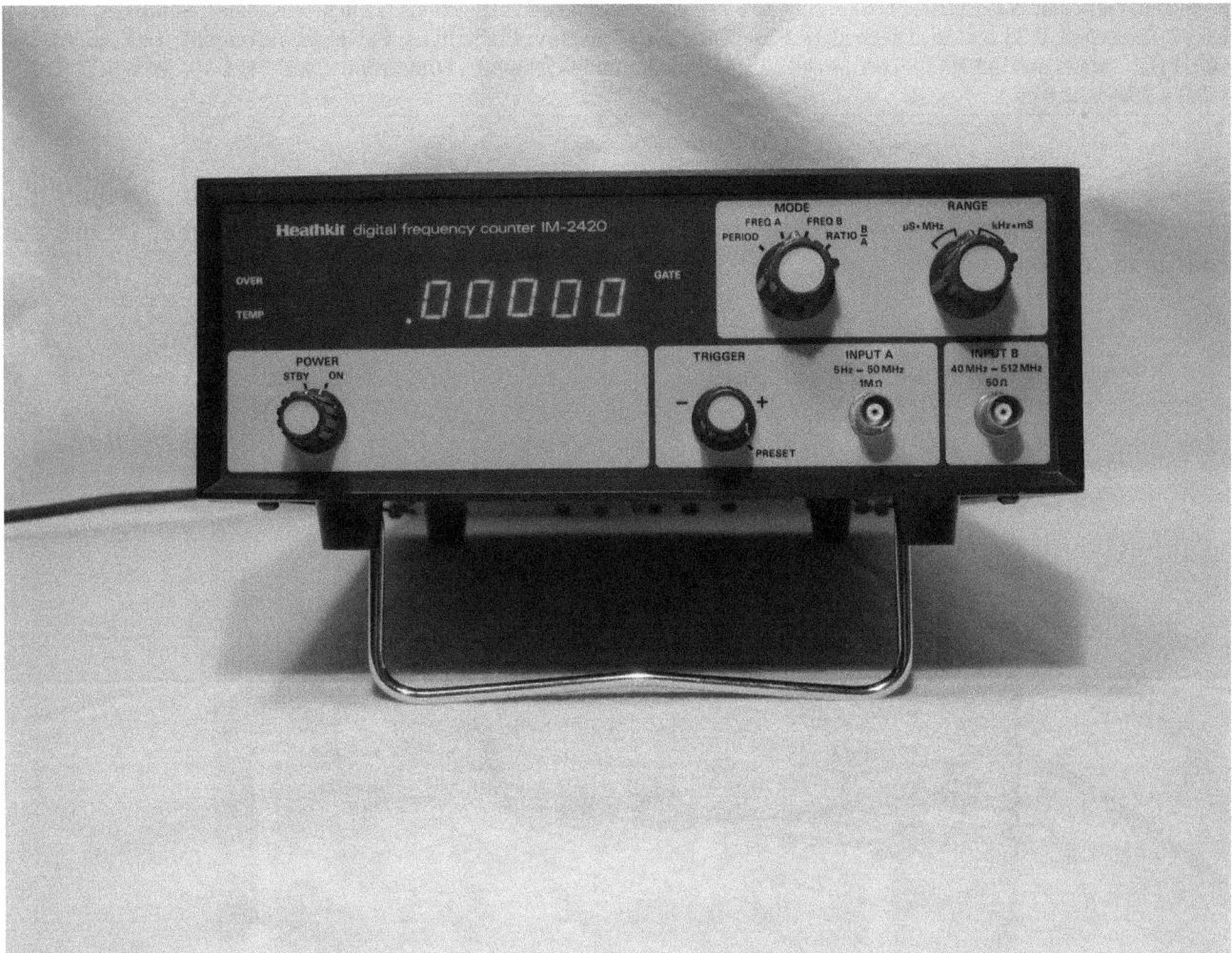

Illustration 11: IM-2420 Front View

It has two inputs. Input A can measure from 5 Hz to 50 MHz and has a one megohm input impedance. Input B covers 40 to 512 MHz and is a 50 ohm input. Full specifications are listed below:

IM-2420 SPECIFICATIONS: Inputs: Frequency Ranges: 5 Hz to 50 MHz, and 40 MHz to 512 MHz. **Sensitivity**: 25 mV RMS guaranteed., 4 to 15 mV RMS typical. **Input Impedance:** 5 Hz to 50 MHz range – 1 megohm shunted by less than 25 pF; 40 MHz to 512 MHz range – 50 ohms nominal. **Input Protection:** 5 Hz to 50 MHz range – 250 VRMS to 100 kHz, derating to 25 VRMS at 50 MHz; 40 MHz to 512 MHz range – 5 VRMS. **Period Measurement Mode: Input:** 5 Hz to 50 MHz only. **Range:** 5 Hz to 10 MHz. **Sensitivity:** 25 mV RMS guaranteed, 4 to 15 mV RMS typical. **Display Resolution (Least Significant Digit):** 100 nS to 0.1 mS, in decade steps. **Ratio B/A Measurement Mode: Input Frequency Limits:** 5 Hz to 25 MHz, and 40 MHz to 512 MHz. **Effective Measurement Range:** From 1.6 to 1.024 x 10(8), guaranteed. **Sensitivity:** 25 mV RMS guaranteed, 4 to 15 mV RMS typical. **Time Base: Frequency:** 10 MHz. **Stability:** To within 0.2 part per million (ppm). **Temperature Stability:** 0.2 ppm from 32 deg. F to 104 deg. F (0 deg. C to 40 deg. C). **Crystal Aging Rate:** Less than 1 ppm per year. **Oven Operating Temperature:** 167 deg. F,+/- 9 deg. F (65 deg. C, +/- 5 deg. C). **Warm-Up Time From Cold Oven (unplugged) Start:** 10 minutes to within 1.0 ppm; 20 minutes to within 0.1

---

ppm. **External Input:** TTL or 2.5 volts RMS from 50 ohm source (10 megahertz). **Input Protection:** Any voltage with peak-to-peak limits between -3.5 volts and +10 volts. **Output:** Will drive 2 standard TTL (i.e. 7400) loads. Short circuit protected. **General: Gate Times:** 0.01 second, 0.1 second, 1 second and 10 seconds (switch-selectable). **Sampling Rate:** Every 0.1 second, 0.33 second, 1 second, or 10 seconds. **LED Display:** Eight digits. **Power Requirement:** 120/240 VAC, 50/60 Hz; 6 watts maximum in STBY mode, 40 watts maximum in ON mode. **Dimensions:** 4.25" H x 10" W x 12.5" D (10.8 x 25.4 x 31.8 cm).

The 0.2 ppm stability is achieved by using a temperature controlled crystal oven. When turned off it is in standby mode where the oven is kept warm to avoid the need to wait for it to warm up (which can take up to 20 minutes).

Illustration 12: IM-2420 Inside View

You can measure the period of channel A, frequency of channel A, frequency of channel B, or the ratio of the frequencies between channels A and B. The mode control selects one of these four modes. A range switch selects one of four ranges for gate times to make the measurement. A trigger control allows adjusting the level at which measurements are triggered. It has a preset mode when turned fully clockwise. The display shows 8 significant digits on red 7-segment LEDs. It has a metal case with rubber feet and a tilt up stand. It runs on AC power only. On the rear panel is a connector that is an input for an external clock source or output of the internal 10 MHz clock, as selected by a switch.

Looking inside you can see it is constructed on a large printed circuit board. It is all solid state using mostly 7400 series integrated circuits. It used a temperature-controlled crystal oven, which is inside a Styrofoam insulated housing. The oscillator runs at 10 MHz.

I bought this unit from a local ham radio operator as part of a lot of Heathkit equipment. When received it was working fine. I generally use it when I need to make accurate or high frequency measurements beyond what the IM-2410 can do.

Both units could be used with an optional telescoping antenna to make readings from a transmitter, for example. While both units provide basic frequency measurements, the IM-2420 sold for over twice the price of the IM-2410. For the higher price you got a maximum frequency of 512 MHz versus 225 MHz for the 2410. You also got a more stable instrument thanks to the temperature controlled crystal oven. The IM-2420 also has two inputs with the ability to measure the frequency ratio between them (to be honest I don't know if this feature is particularly useful.) I don't have original manuals for either unit but partial manuals and schematics are available on the Internet and full manuals can be purchased from a number of suppliers.

Compared to older instruments for frequency measurement like grid dip meters, digital frequency counters are considerably more accurate and easier to use. Both of these units were a good value at the time, in part because of the cost savings of offering them in kit form.

Chapter 6: Meters

We start this section with an introduction to the theory behind meters and the different types available, before moving on to discuss Heathkit's offering.

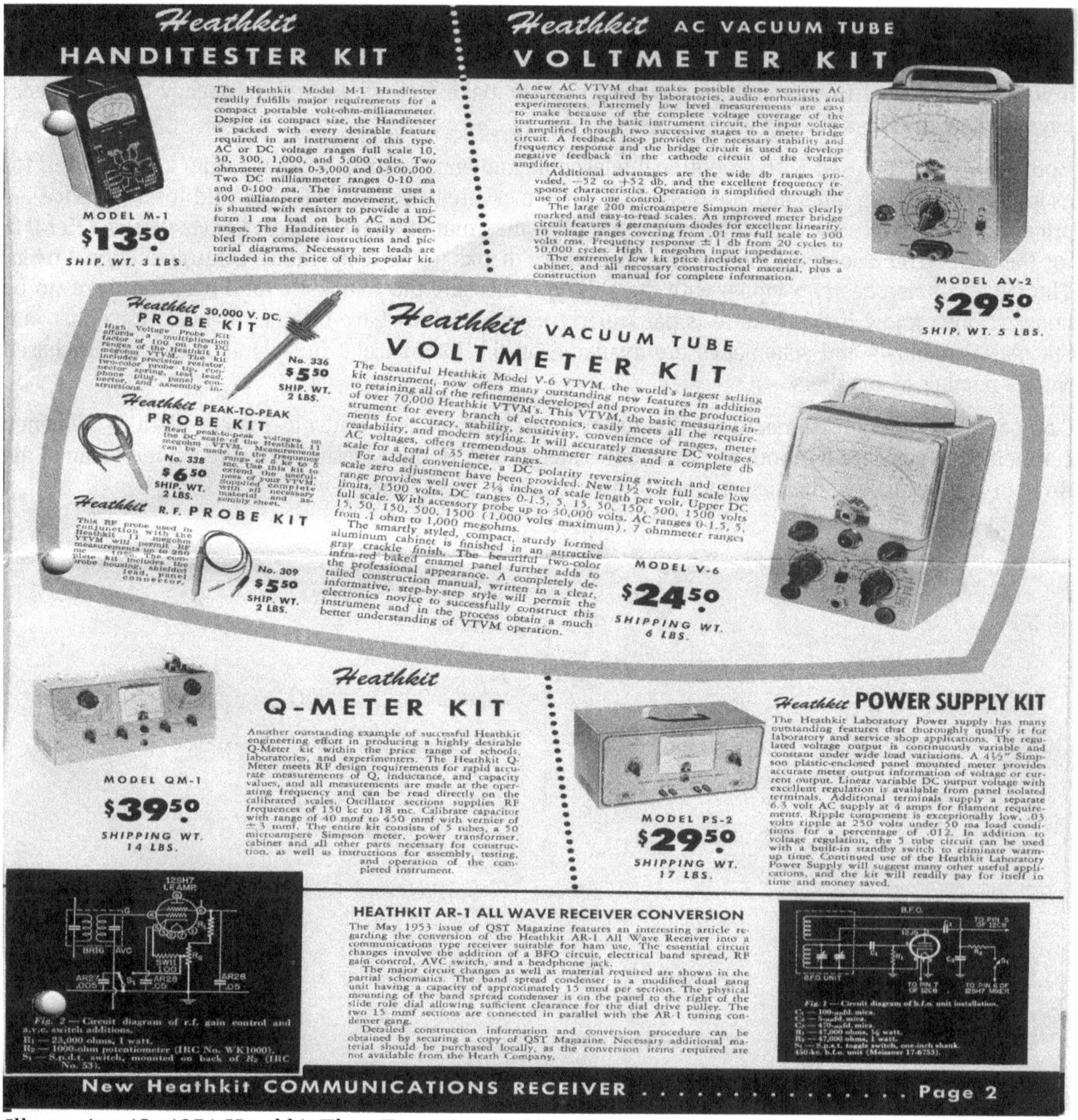

Illustration 13: 1951 Heathkit Flyer Featuring Various Meters

The first practical meter for measuring electrical current was developed by Jacques-Arsène d'Arsonval in 1882 and subsequently improved by Edward Weston. Known as the *D'Arsonval Movement*, it used a pivoting indicator needle and spring, permanent magnet, and small coil to indicate electrical current. A modern movement can display a full scale reading for current as small as 50 microamps. By adding a large series resistor to the meter, it can measure voltage of any desired range. Placing a small *shunt resistance* across the meter allows it to measure larger currents. With a small power source, such as a battery, and a resistor adjusted to read full scale when the circuit is closed, the meter can be used to measure resistance. Illustration 14 shows some simplified circuits illustrating how this is done for a hypothetical meter using a 1 mA movement.

Volt-Ohm-Milliammeter (VOM)

Meter: 0-1mA, 1Ω internal resistance

A - Voltage Measurement	B - Current Measurement	C - Resistance Measurement

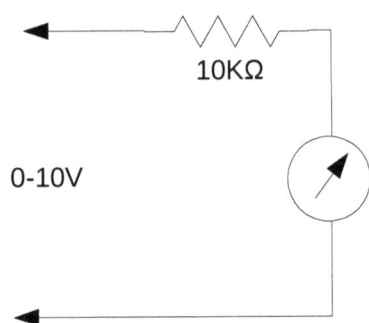

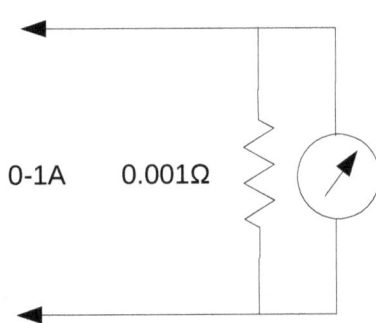

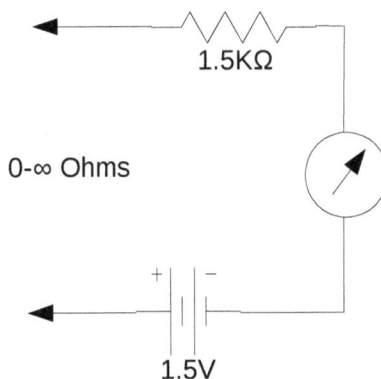

Illustration 14: Volt-Ohm-Milliammeter Circuits

Let's analyze the circuit in Illustration 14. If you are familiar with Ohm's law relating voltage, current, and resistance, you will be able to easily derive the numbers used in the calculations. If not, you can just accept the numbers used and follow the concepts.

Referring to circuit A in Drawing 1, if we put a 10 kilohm resistor in series with the meter, an applied input voltage of 10 volts will cause 1 milliamp of current to flow, causing the meter to read full scale. This gives us a voltmeter with a 0 to 10 volt range. By using various values of series resistors, ideally selected by a switch, we can provide different voltage ranges. In circuit B, a small shunt resistor is placed in parallel with the meter. If a current of one ampere was flowing through the input terminals, 999 milliamps would flow through the shunt resistor and one milliamp would flow through the meter. Note that here we have to take into account the 1 ohm internal resistance of the meter itself. This implements a current meter with a full scale reading of one ampere. By selecting different values of shunt resistance we can support other current ranges. In circuit C we connect a 1.5 volt battery and a 1.5 kilohm resistor in series with the meter. If we connect the test leads directly together, one milliamp of current will flow and the meter will read full scale. With the test leads shorted this is zero ohms and we could mark full scale on our meter as zero. If the meters leads are kept open, no current will flow and the meter will not deflect. We can mark this far left position on the meter as infinite resistance (often indicated using the symbol ∞). If we placed a 1.5 kilohm resistor across the test leads a current of 0.5 milliamps would flow and the meter would read half-scale. We could mark this as 1500 ohms.

Similarly, we could mark the meter scale with various intermediate values so that we have a practical ohmmeter than can measure resistance.

You can see that with a suitable switching arrangement one could implement a VOM that measured voltage, current, and resistance with different ranges using the three basic circuits above.

Volt-Ohm-Milliammeters (VOMs)

A device that can measure voltage, resistance, and current is known as a Volt-Ohm-Milliammeter, Volt-Ohm Meter or VOM, and is one of the most commonly used pieces of electronic test equipment. Units typically have a switch that allows selecting different ranges for voltage, current, and resistance. Most units provide the facility for measuring AC voltage as well. Most are powered by internal batteries (for the ohms function) so they are self-contained and portable.

VOMs need to draw some current from the circuit under test to deflect the meter. A unit with a 1 mA meter will require 1 mA for a full scale reading. This important aspect of a VOM is the sensitivity rating usually specified as the inverse of current, in ohms per volt. A unit with a 1 mA meter, for example, would be a 1,000 ohms per volt meter. If it used a 50 µA meter movement, it would be rated at 20,000 ohms per volt. This measurement unit makes is easier to quickly calculate the resistance of the meter on a given range. For example, if a 1,000 ohms per volt meter was on the 10 Volt range, its input resistance would be 10,000 Ohms.

If the resistance of a meter is high compared to the values in the circuit being measured, the current drawn by the meter will have a negligible effect on the circuit. But if the meter resistance is comparable to or lower than those in the circuit, the current drawn can have a significant loading effect on the circuit, causing the readings to be inaccurate, or in the worst case causing the circuit to no longer function. A meter with a higher ohms per volt rating is therefore desirable. Service manuals for equipment like old radios will sometimes specify the ohms per volt rating of the meter used to make measurements so that the expected readings published in the manual will match what is measured when the meter is loading the circuit.

Heathkit made over a dozen different models of VOM. A typical early model was the MM-1, a 20,000 ohms per volt meter with seven voltage ranges, five current ranges, and three resistance ranges. During the time that Heathkit was owned by Daystrom they offered some meters made by their partner company, Weston, like the EUP-26 which was a Weston 980 meter. An interesting VOM was part of the EK-1 educational course on electricity and electronics. As part of the course the student built a VOM and used it in experiments to learn concepts such as Ohm's law and series and parallel circuits. Table 8 lists the models of VOM that Heathkit offered.

Table 8: Heathkit VOMs

Model	Description	First Year	Last Year	Comments
EUP-26	VOM	1968	1970	Weston Meter
IM-16	VOM	1967	1974	Bench type, 6" meter, solid-state
IM-17	VOM	1967	1977	High impedance FET, battery-powered, in storage case
IM-25	VOM	1967	1974	Bench type, 6" meter, solid-state, high-impedance
IM-104	VOM	1973	1976	FET VOM, 4.5" meter
IM-105	VOM	1971	1981	Taut band 4.5" meter
IM-1104	VOM	1979	1981	10 MΩ input, portable
IM-5217	VOM	1979	1987	Comes with case

Model	Description	First Year	Last Year	Comments
IM-5284	VOM	1977	1983	Solid-state, battery or AC power
M-1	VOM	1950	1962	Pocket-sized "Handitester"
MM-1	VOM	1951	1967	5", 20KΩ per volt, Battery operated portable
SM-660	VOM	1971	1981	Assembled version of IM-105
SM-666	VOM	1973	1976	Assembled version of IM-104

Vacuum-Tube Voltmeters (VTVMs)

An improvement over the VOM is the Vacuum-Tube Voltmeter, or VTVM. It uses a bridge circuit and a vacuum tube as an amplifier. With this design the meter does not need to be powered by the circuit under test, allowing the input resistance to be much higher. The typical VTVM has an input resistance that is fixed at 10 or 11 Megohms on all ranges. VTVMs can also be more sensitive since they amplify the input signal, and can use a larger meter. The disadvantages of VTVMs are larger size, cost, and the need to be run on AC power. As equipment moved to solid-state, VTVMs were replaced by designs that typically used Field Effect Transistors (FETs). These offered similar functionality to VTVMs but could be battery powered.

The V-1 VTVM was one of the first Heathkits to come on the market in 1947, just after the O-1 oscilloscope. The circuitry is very similar in most of the Heathkit VTVMs; some only differed in the style of case and knobs. The V-7 was the first with the components mounted on a printed circuit board rather than point to point wiring. A popular VTVM was the IM-18 which will be covered in detail later in this chapter.

Most of the models used a test probe with an internal one Megohm resistor that was switched in for DC voltage measurements and out for measuring resistance and AC voltage. If you purchase a used unit, make sure you get the probe. It is possible make one but don't be tempted to omit it as it is required for proper operation. Most VTVMs used an internal battery for the ohms function which may have leaked if left in the unit for a long period of time. It is possible to replace it with a small power supply, if desired.

While most VTVMs could measure AC voltage, Heathkit also sold some dedicated AC voltmeters that could accurately measure AC RMS voltages down to 0.01 volts and were often used for measuring audio frequency signals.

Heathkit later offered some FET meters that were comparable in features to VTVMs but solid-state. An interesting transition product was the IMA-18-1. It consisted of two plug-in modules that replaced the 6AL5 and 12AU7 tubes in many VTVMs like the IM-18 and IM-28, converting them to solid-state. This avoided the need to wait for the tubes to warm up and reduced pointer drift corrections.

The fact that the IM-5228 VTVM was still being sold in 1989, long after most test equipment had moved entirely to solid-state, is a testament to the popularity of Heathkit's VTVMs.

Table 9: Heathkit VTVMs

Model	Description	First Year	Last Year	Comments
AV-1	VTVM	1951	1951	10 ranges
AV-2	VTVM - AC	1951	1956	0.01V RMS to 300V RMS
AV-3	VTVM - AC	1957	1961	10 ranges
EF-1	VTVM Course			For IM-18, IM-28, etc. Includes power supply.
EUW-24	VTVM	1963	1970	4.5", 7 ranges, assembled
IGW-18	VTVM			Berkeley Physics Laboratory, assembled
IM-10	VTVM	1960	1962	Deluxe model, 6" meter, 7 ranges
IM-11	VTVM	1961	1968	Replaced V-7A
IM-13	VTVM	1963	1968	Bench type, 6" meter
IM-18	VTVM	1968	1976	4.5", 7 ranges, probe
IM-21	VTVM - AC	1961	1968	AC RMS voltage, 0.01 to 300 VAC in 10 ranges
IM-28	VTVM	1968	1976	Bench type, 6" meter
IM-32	VTVM	1962	1963	Replaced IM-10
IM-38	VTVM - AC	1968	1976	AC RMS voltage, 0.01 to 300 VAC in 10 ranges
IM-5218	VTVM	1977	1983	4.5", 7 ranges, assembled
IM-5228	VTVM	1977	1989	Bench type version of IM-5218, 6" meter
IM-5238	VTVM - AC	1976	1981	1 mV to 300 V AC, 12 ranges
IMW-18	VTVM			Assembled version of IM-18
SM-20A	VTVM			Assembled version of IM-18
SM-21A	VTVM			Bench type, assembled
SM-22A	VTVM - AC			Assembled version of IM-38
SM-600	Digital VTVM			No details known
V-1	VTVM	1947	1948	5 ranges
V-2	VTVM	1948	1949	6 ranges
V-3	VTVM	1949	1949	Battery-operated, engineering problems, never produced
V-4	VTVM	1950	1950	6 ranges
V-4A	VTVM	1950	1950	6 ranges
V-5	VTVM	1951	1951	6 ranges
V-5A	VTVM	1951	1951	6 ranges
V-6	VTVM	1951	1954	7 ranges
V-7	VTVM	1955	1955	7 ranges, First to use PCB
V-7A	VTVM	1956	1961	4.5" meter, 7 ranges

Digital Multimeters (DMMs)

By the 1970s, digital electronics made it feasible to make measurements and directly display them as digits using LEDs or LCDs. These Digital Multimeters (DMMs) are standard today and low cost units can accurately measure voltage, current, resistance and often also other units such as temperature, frequency, and capacitance. Compared to VTVMs they are more accurate, smaller, and easier to use.

Many automatically switch ranges and have protection from overvoltage and overcurrent. Analog meters still exist because they have one advantage: when adjusting circuits for a peak or a null it is much easier with an analog meter than with a digital indicator. For this reason some DMMs also provide an analog indication in the form of a bar graph. A typical DMM offered by Heathkit was the IM-2215 which we will cover in more detail later in this chapter. Table 10 lists the offering of Heathkit DMMs.

Table 10: Heathkit DMMs

Model	Description	First Year	Last Year	Comments
ETI-7010	DMM	1990	1990	Bench type, assembled
ID-2311	DMM	1989	1991	3.5 digit
IM-102	DMM	1971	1978	3.5 digit, Nixie display, 26 ranges
IM-1202	DMM	1973	1975	2.5 digit, Nixie display
IM-1210	DMM	1977	1981	2.5 digit, LED display
IM-1212	DMM	1975	1976	2.5 digit, Nixie display
IM-2202	DMM	1975	1981	3.5 digit, LED display
IM-2212	DMM	1979	1981	3.5 digit, LED display, auto-ranging
IM-2215	DMM	1979	1984	3.5 digit, LCD display, hand-held
IM-2260	DMM	1982	1991	3.5 digit, LED display
IM-2262	DMM	1982	1983	3.5 digit, LCD display, true RMS
IM-2264	DMM	1982	1987	3.5 digit, LCD display, true RMS, crest factor indicator
IM-2320	DMM	1987	1990	3.5 digit, LCD display, hand-held
SM-1210	DMM			3 digit, LED
SM-1212	DMM	1975	1976	Assembled version of IM-1212
SM-2206	DMM	1983	1986	Clamp-on multimeter
SM-2208	DMM	1988	1990	Clamp-on ammeter, 300A
SM-2215	DMM	1979	1984	Assembled version of IM-2215
SM-2255	DMM	1988	1990	3.5 digit, hand-held
SM-2260	DMM	1982	1991	Assembled version of IM-2260
SM-2300	DMM	1986	1988	Shirt pocket sized, assembled
SM-2300A	DMM	1989	1991	Shirt pocket sized, assembled
SM-2310A	DMM	1989	1989	Shirt pocket sized, assembled
SM-2311	DMM	1989	1992	Assembled version of ID-2311
SM-2320	DMM	1986	1986	Heavy duty portable, assembled
SM-2360	DMM	1989	1990	Bench/portable, assembled
SM-2372	DMM/Frequency Meter	1989	1992	20 MHz, assembled
SM-2374	DMM	1989	1992	Clamp-on, 1000A, AC/DC volts
SM-2380	DMM	1990	1992	Assembled, Heath branded

Dip Meters

A *Dip Meter*, also called a grid dip meter, grid dip oscillator, or just dipper, is an instrument used to measure the resonant frequency of tuned circuits. It is an oscillator whose output energy changes in the vicinity of a resonant circuit which is tuned to the frequency at which the oscillator generates. Grid dip meters were first developed in the 1920s and were built with vacuum tubes. The devices measured the value of the tube's grid current on a meter, which dipped when tuned to resonance, hence the name. Later meters used solid-state devices (which don't have a grid element) and are just called dip meters. Dip meters have been widely used by amateur radio operators for measuring the properties of resonant circuits, filters, and antennas. Applications include:

- measuring the resonant frequency of a tuned circuit,
- measuring the frequency of an oscillator,
- acting as a signal source for adjusting radio circuits,
- measuring inductance, capacitance, and quality factor (Q),
- acting as a field strength meter, and
- testing antennas and transmission lines.

Heathkit sold a number of dip meters over the years. The first was the GD-1, coming on the market in 1951. It was replaced in 1952 by the slightly improved GD-1A, which in turn was replaced by the further improved GD-1B in 1954. The GD-1B was produced until 1960. In 1961 the HM-10 Tunnel Dipper was introduced. It used the relatively uncommon semiconductor tunnel diode rather than a tube. The HM-10A replaced it in 1962 and was produced until 1970. The Solid State HD-1250 Dip Meter came on the market in 1975 and was produced until 1991. Most units used about seven changeable coils to cover the radio frequency range, typically 1.6 to 250 MHz. A dial scale indicates frequency. A level control adjusts the output level of the oscillator, and a sensitive meter indicates the dip or peak in signal level during use. The dip meter seems to have been a neglected piece of test equipment and historically not many people knew what it was used for. In his book *Tube Testers and Classic Electronic Test Gear* Alan Douglas says the dip meter is the "single most versatile test instrument for RF use". Any single function can probably be better performed by a specialized instrument, whether it is a frequency generator, frequency counter, or LC meter. But no one other low cost instrument can be used for so many functions. In fact, dip meters are still being manufactured today, little changed from the designs from the 1940s, other than using solid-state components. Table 11 list the models of dip meters offered by Heathkit over the years.

Table 11: Heathkit Dip Meters

Model	Description	First Year	Last Year	Comments
GD-1	Grid Dip Meter	1951	1951	With plug-in coils
GD-1A	Grid Dip Meter	1953	1953	With plug-in coils
GD-1B	Grid Dip Meter	1953	1960	With plug-in coils
HD-1250	Dip Meter	1975	1991	Solid state, 1.6 MHz to 250 MHz, 7 plug-in coils
HM-10	Tunnel Dipper	1961	1962	Solid state (tunnel diode), plug-in coils
HM-10A	Tunnel Dipper	1962	1970	Solid state (tunnel diode), plug-in coils

Miscellaneous Meters

A *clamp-on meter* or clamp-on probe for a meter is a device that wraps around an electrical conductor and allows AC current to be measured without breaking the circuit or having to make physical contact with it. These are best suited for measuring high currents. Heathkit sold a few meters of this type. Typically they measured large AC currents with ranges to 6 amps or more. Some models also offered

AC voltage measurement, using test leads rather than the clamp which was used only for current measurement.

For measuring the high voltages used with televisions CRTs, Heathkit offered adaptors that allowed a standard VTVM to safely measure high voltages. They also offered dedicated high voltage meters like the IM-5210 that could directly measure up to 40,000 volts.

While the previously mentioned capacitor testers and impedance bridges can measure capacitance, Heathkit sold some dedicated capacitance meters. The early model, the CM-1, displayed the value on an analog meter. The IT-2250, offered in the 1980s, was a digital meter with an LCD display, similar to a DMM.

A Distortion Meter can measure harmonic distortion, an indication of how accurately an audio signal is being reproduced. They are primarily used for testing audio circuits like high fidelity or stereo audio systems. Heathkit offered a number of models. The IM-48 and IM-58 were two typical models.

Power meters can measure the power level of a signal. Some units are designed to measure power of audio signals for consumer audio systems. Others measure power of radio frequency (RF) signals, and are often used by amateur radio operators.

Finally, one oddity under the category of meters was the IM-103 voltage monitor. It could perform only one function – display the AC line voltage on a large analog meter. While it was marketed as being useful in electronic labs where the line voltage might affect instrument calibration, it also appealed to radio amateurs and other electronics hobbyists. Table 12 lists the miscellaneous meters made by Heathkit over the years.

Table 12: Heathkit Miscellaneous Meters and Accessories

Model	Description	First Year	Last Year	Comments
309	Probe - RF	1952		For use with V-1 VTVM
310	Probe - HV			10 kV
336	Probe - HV	1952		30 kV for use with VTVMs
309-B	Probe - RF			RF up to 250 MHz, for use with V-1
309-C	Probe - RF			For use with VTVM, scope, signal tracer
337-B	Probe - RF Demodulator			For use with VTVM, scope, signal tracer
338-B	Probe - Peak-to-Peak			5 Hz to 5 MHz
AF-1	Audio Frequency Meter	1951	1959	20 Hz to 100 kHz
CM-1	Capacitance Meter	1956	1960	Analog meter
HD-1	Harmonic Distortion Meter	1958		Measures voltage and % distortion
IM-1	Distortion Analyzer	1951	1954	Measures IMD, full scale ranges of 30%, 10%, 3%
IM-12	Harmonic Distortion Meter	1963	1967	Analog meter
IM-48	Audio Intermodulation Analyzer		1976	AC VTVM, Wattmeter, distortion analyzer
IM-58	Harmonic Distortion Meter		1976	Measures noise and distortion, 20 to 20 kHz
IM-103	Line Voltage Monitor	1970	1976	90 to 140 VAC
IM-4180	FM Deviation Meter	1979	1987	For FM transmitter testing
IM-4190	Bi-Directional Wattmeter	1978	1981	Forward and reflected power, 100 to 1000 MHz, 300W

Model	Description	First Year	Last Year	Comments
IM-5210	Probe Meter	1975	1984	40 kV HV probe meter
IM-5215	Probe Meter	1984	1991	40 kV HV probe meter
IM-5225	Multimeter	1977	1981	FET analog
IM-5248	Distortion Analyzer	1976	1979	Distortion 0.1 to 100% in 3 ranges, AC voltmeter 10 mV to 100 V
IM-5258	Distortion Analyzer	1976	1983	Distortion 0.03 to 100% in 6 ranges, AC voltmeter 1 mV to 300 V
IMA-18-1	Solid State Tube Replacements			Replaces tubes to convert IM-18, IM-28 and other VTVMs to solid state
IMA-100-10	Probe - HV		1987	X100 30 kV HV probe for 10 MΩ meters
IMA-100-11	Probe - HV	1985	1986	X100 30 kV HV probe for 11 MΩ meters
IMA-1000-1	Probe - HV			X1000 HV probe for 1 MΩ meters
IMA-2215-1	Case			Leather case for IM-2215or IT-2250
IT-2250	Capacitance Meter	1981	1987	Digital hand-held
PK-3	Probe - RF			100 MHz, for VTVM
PK-3A	Probe - RF			For DC voltmeters
PKW-4	Probe - for IM-5218/IM-5228			VTVM replacement probe
PKW-200	Test Lead Set	1981		Shielded red/black cables with probes
SM-4180	Frequency Deviation Meter	1979	1987	Assembled version of IM-4180
SM-5210	Probe Meter	1975	1984	Assembled version of IM-5210

In-Depth: The IM-18 VTVM

Let's take detailed look at a typical model of VTVM offered by Heathkit, the IM-18. Illustration 15 shows the front panel of the unit.

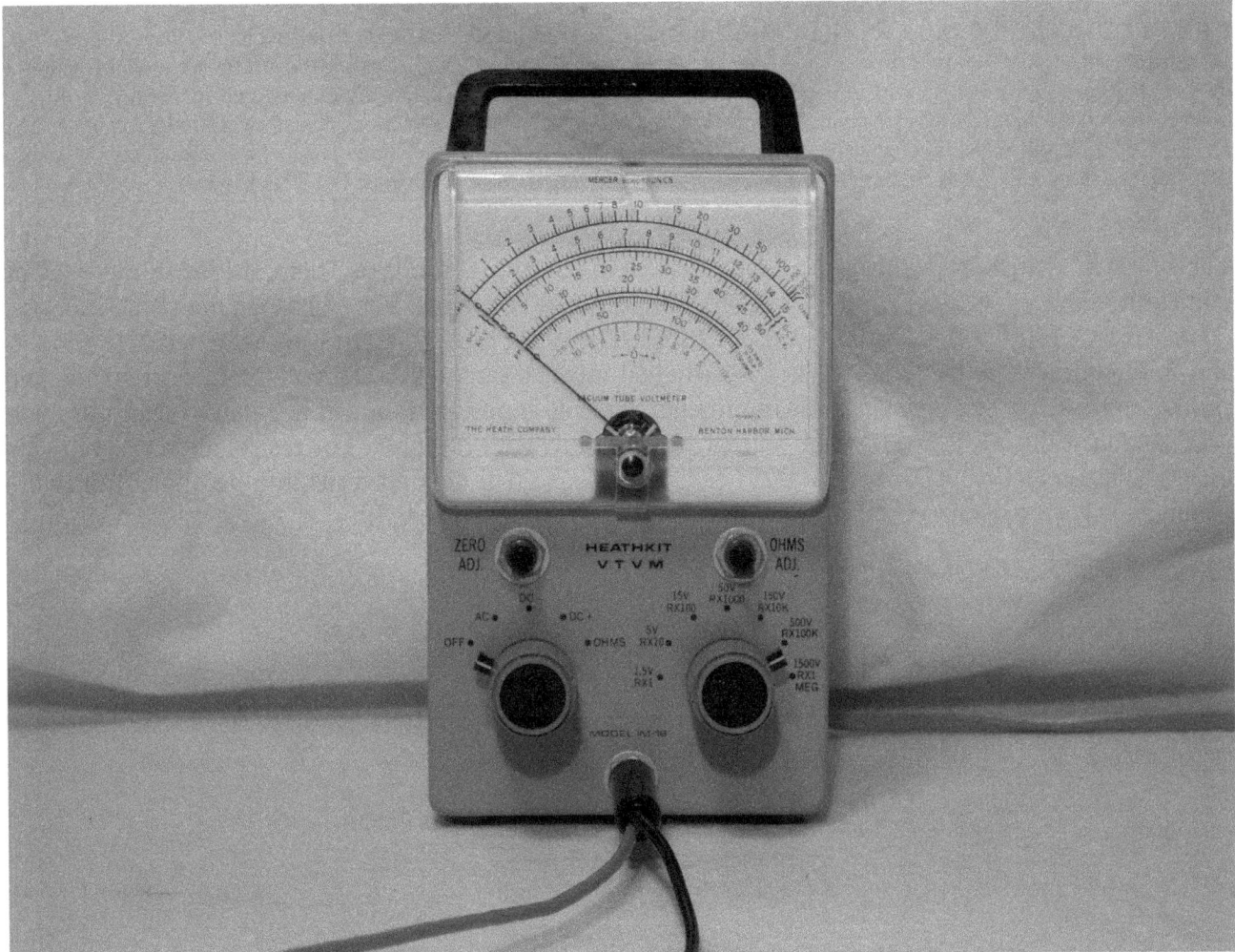

Illustration 15: IM-18 Front View

Like most Heathkits, it was sold as a kit. It came on the market in 1968 and was produced until 1976. It was the successor to the IM-11 that came on the market in 1961, that in turn was the successor to the V-7 and V-7A. There are only minor differences between the IM-11 and the IM-18, mostly a new look and color. In 1977 the IM-18 was replaced by the IM-5218 and was produced until 1983. Again, it was only a new look and color. The VTVM V-1 was one of the first Heathkits, coming on the market in 1947, just after the oscilloscope O-1. The circuitry is very similar in all the models.

The IM-18 is a typical VTVM and can measure AC and DC voltage and resistance over seven ranges. It features a large 4-½ inch analog meter. It runs from the AC line although it also uses an internal battery for the resistance function.

A 1971 Heathkit catalog lists it at $29.95 with the factory wired IMW-18 model at $49.95. In a 1976 catalog the price was $36.95 and the wired version, now designated as the SM-20A, was $70.00. It was described as "our most popular VTVM". The published specifications are as follows:

IM-18 SPECIFICATIONS: Meter scales: DC & AC (rms): 0-1.5, 15, 50, 150, 500, 1500 V full scale. AC peak-to-peak: 0.4, 14, 40, 140, 400, 1400, 4000 V full scale. Resistance: 10 ohm center scale x1, x10, x100, x10k, x100k, x1 meg. Measures 1 ohm to 100 megohms. Meter: 4½" 200 µA movement: Input resistance: 11 megohms (1 megohm in probe) on all DC ranges: 1 megohm shunted by 35 pF on all AC ranges. Circuit: Balanced bridge (push-pull) using twin triode. Accuracy: DC ±3%, AC ±5% of full scale. Frequency response ±1 dB, 25 Hz to 1 MHz (600 ohm source). Battery requirements: 1.5V, size "C" cell (not supplied). Power requirement: 120/240 VAC, 60/50 Hz, 10 W. Dimensions: 7-3/8" H x 4-11/16" W x 4-1/4" D.

The function switch selects the mode, either AC, DC- or DC+, or ohms. There are seven ranges. For voltage it offers 1.5V, 5V, 15V, 50V, 150V, 500V, and 1500V ranges. For resistance the ranges are X1, X10, X100, X1000, X10K, X100K, and X1 MEG. There is a probe that must be switched between DC and AC/Ohms. A one megohm resistor is switched into the circuit for DC voltage measurements and shorted out for AC and resistance measurements. After it is powered on and allowed to warm up, you need to adjust the ZERO ADJ. knob so that it reads zero with no input. For resistance measurements there is also an OHMS ADJ. control that needs to be set so that the unit reads infinite ohms (full scale) when the leads are open.

Illustration 16: IM-18 Inside view

Illustration 16 shows the inside of the unit. It uses two tubes, a 6AL5 duo diode (for rectifying AC input signals) and a 12AU7 dual triode (for the bridge circuit). The power supply uses a solid state diode. It requires AC power as well as a 1.5 V "C" battery for the ohms function.

The meter has an 11 megohm input impedance, higher than some modern DMMs. It can measure quite high resistances with ranges up to R X1 MEG. Because one of the test leads is connected to ground, it can only measure voltages referenced to ground. Some people have modified their units to have a floating input.

Calibration of the unit can be done without instruments. DC calibration is done against the 1.5 volt battery. Resistance does not require any calibration. AC voltage can be calibrated against the AC line voltage or against a known AC voltage.

I bought this unit used on eBay, mostly for nostalgia reasons. In my teens I was given an IM-13, a similar but slightly larger bench model. I used it for my early forays into electronics and radio.

The meter is not very accurate by today's standards. The accuracy is ±3% of full scale for DC and ±5% for AC. It is difficult to read accurately as compared to a DMM. Where most modern DMMs are autoranging, it requires manually changing ranges. It also has limited functions: it can't measure capacitance or even current. The switch in the test lead is easy to forget to set correctly and sometimes breaks after a few years of use. In fact, the test leads are often missing in used models that come up for sale. Nevertheless, while not commonly used for working on modern electronics, the Heathkit IM-18 VTVM is a classic piece of test equipment. Because it is a Heathkit, if it fails you can probably repair and calibrate it yourself because of the availability of a manual and parts. Many thousands of them were sold and used for years by hobbyists and repairmen and many are still in operating condition and used occasionally like this one.

In-Depth: The IM-2215 Digital Multimeter

Now let's take a detailed look at a late model DMM offered by Heathkit, the IM-2215.

This is a portable, battery-operated, hand-held Digital Multimeter that was offered from 1979 through 1984. It retailed for $94.95 in 1979 (roughly equivalent to $300 in today's dollars). A factory assembled and calibrated version, the SM-2215, was also available at higher cost. It features a 3-½ digit LCD display and can measure resistance and AC and DC voltage and current. It is not auto-ranging, but does have but auto polarity, auto decimal point, and leading zero suppression.

The specifications, taken straight from a Heathkit catalog, are as follows:

IM/SM-2215 SPECIFICATIONS: Voltage: Ranges: DC, 200 mV, 2 V, 20 V, 200 V, 1000 V; AC 200 mV, 2 V, 2 0V, 200 V, 750V (rms). **Accuracy: Lab Standards:** DC +/- 0.25% of reading+ 1 count SM-2215, +/- 0.1%); , AC +- 0.5% of reading+ 3 counts. **Built-in Standards:** DC, +/- 0.35% or reading + 1 count; AC +/- 0.6% or reading + 3 counts. **Input impedance:** 10 Megohms (shunted on AC by approximately 100pF). **Frequency Response for AC Measurements (25 deg.C, +10 deg.C):** On 200 mV, 2V, 20 V and 200 V ranges, 40 Hz-1 kHz; on 750 V range, 40 Hz-450 Hz (both specifications are +/- 1% or reading + 3 counts). **Current Ranges:** DC, 2 mA, 20 mA, 200 mA; AC, 2 mA, 20 mA, 200 mA, 2000 mA. **Accuracy:** DC +/0 0.75% of reading + 1 count (SM-2215, +/0 -.35%); AC, 2 mA range, +/- 1.5% or reading + 3 counts, 40-200 Hz; other AC ranges, +/- 1.5% or reading + 3 counts, 40 Hz-1kHz. **Resistance Ranges:** 200, 2K, 20K, 200K, 2M, 20M, +/- 0.25% of reading + 1 count (SM-2215, 2K-2M: +/- 0.15%), 20 Megohms range, +/- 2% or reading + 1 count.

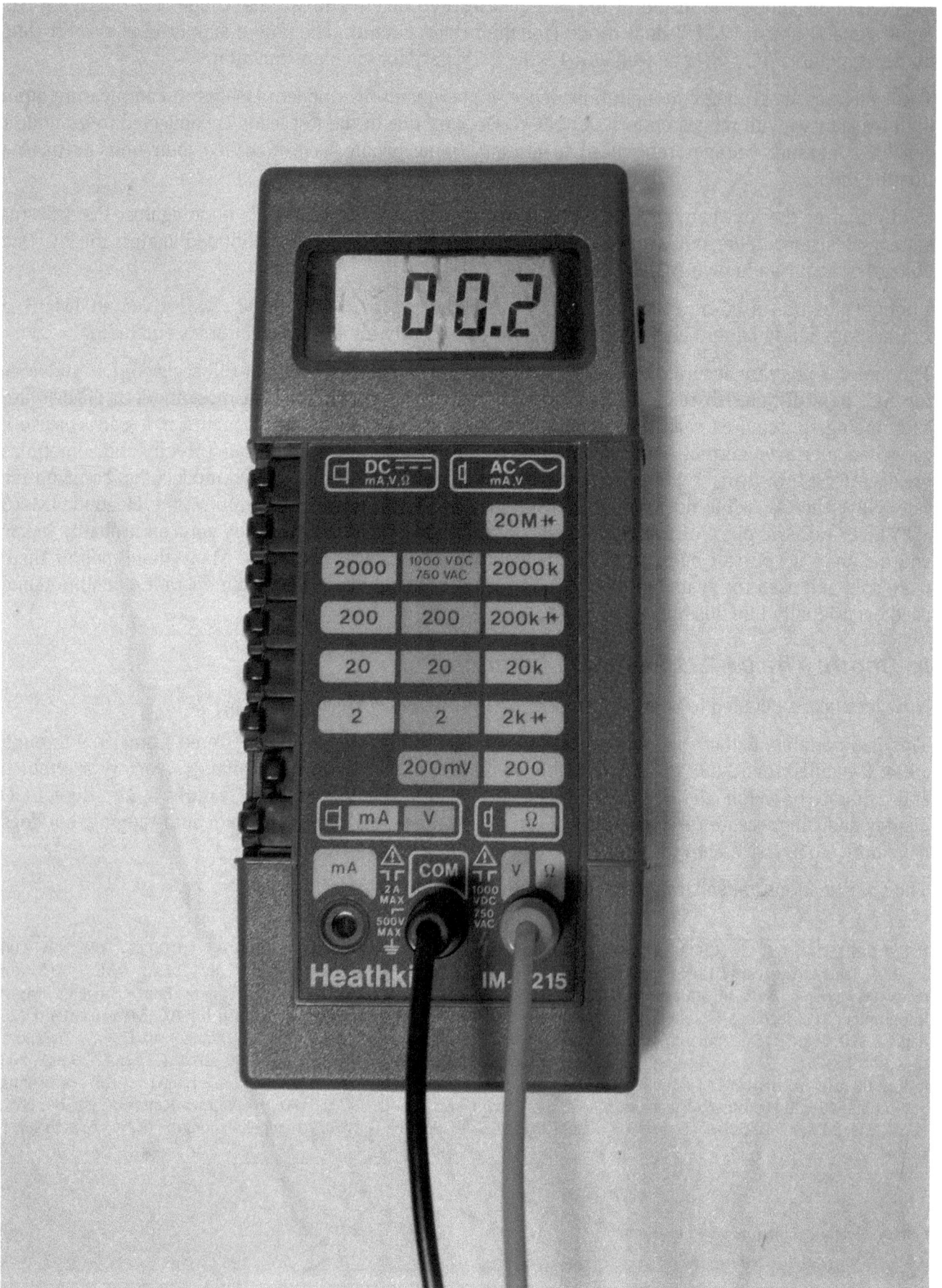

Illustration 17: IM-2215 Front View

Illustration 17 shows the front view of the unit in operation. As well as a power switch, controls include eight pushbutton switches to select the function and range. The unit can be operated while held in one hand. There are three banana type jacks: a common lead and inputs for current and voltage/resistance. It is fuse and diode protected against overloads.

The LCD display can indicate overrange and low battery conditions. It can operate for up to 200 hours on a standard 9 volt battery. An optional external power adaptor was also available. It came with a set of test leads and the case has a built in pivoting stand. There was an optional leather case and it could be used with Heathkit's high voltage and RF probes. The blue color case is similar in style to Heathkit's other instruments of the time.

Illustration 18 shows a view inside the unit with the front cover removed. Circuity is on a single printed circuit board and includes five transistors, four integrated circuits, an LCD display module, and some discrete components. Most of the functionality is handled by an ICL7106, a 40 pin MOS LSI digital multimeter chip that incorporates dual slope A/D conversion and the LCD display driver circuitry.

Illustration 18: IM-2215 Inside View

To simplify calibration, the meter includes an internal voltage reference that it can be calibrated against without any instruments. It can also be calibrated against an external reference if one is available, for improved accuracy. The internal calibration procedure is summarized on a label inside the meter.

A 1982 Canadian Heathkit catalog lists the IM-2215 at $199.95 and the factory assembled SM-2215 at $279.95. The AC adaptor was $9.95 and the leather case was $29.95.

I picked up the unit shown here at a garage sale in 2009 for around $5, a deal I couldn't pass up. It came with test leads but no manual, however a partial manual is available on the Internet.

The troubleshooting chapter of the manual has an interesting section called "Circuit Board Cleaning" that I quote here:

Use the following "last resort" procedure to clean a contaminated main circuit board.
1. Remove the bezel, the liquid crystal display, and the LCD holder from the display circuit board.
2. Use demineralized water and a soft brush to clean the entire circuit board and the pushbutton switch assembly. CAUTION: Avoid getting excessive amounts of water in the switches.
3. Bake at 150°F for 5 hours. CAUTION: Allow the circuit board time to cool down before you reassemble the instrument.

I can't vouch for whether this procedure is applicable to other equipment, and personally I would indeed only use it as a last resort!

In summary, the IM-2215 is quite comparable to today's DMMs. It lacks auto-ranging and facilities like measuring capacitance that many modern meters offer. There are also modern DMMs that are much lower in cost than this unit was, although I would rate it as more rugged and better protected than some of the low-cost units on the market today. Because the calibration procedure is documented in the manual, this unit can also be kept accurate even though it is over 30 years old.

In-Depth: The HD-1250 Dip Meter

In this section we'll take a look at the the HD-1250 Solid-State Dip Meter. The last in a series of dip meters, it came on the market in 1975 and was produced until 1991. A 1977 Heathkit catalog lists it for a price of $59.95 and says you can "build it in one evening". My 1982 Canadian Heathkit catalog lists it at a price of $129.95.

The HD-1250 is similar to many dip meters offered by different manufacturers over the years. It uses seven changeable coils to cover the range of 1.6 to 250 MHz. A dial scale indicates frequency. A level control adjusts the output level of the oscillator, and a sensitive 150 microamp meter indicates the dip or peak in signal level during use. It uses two transistors and runs from a 9 volt battery. A headphone jack is provided which can be used to listen in to a signal which is modulated. The published specifications of the HD-150 are given below:

SPECIFICATIONS

Frequency Range	1.6 to 250 MHz.
Controls	Tuning capacitor.
	Oscillator level control.
	On/Off switch.
Meter Movement	150 microampere.
Solid-state Circuits	1 NPN transistor oscillator.

---

	1 Dual-gate MOSFET amplifier.
	2 Diffused silicon hot carrier diode detectors.
Power Source	9-volt NEDA Type 1604 battery.
Dimensions (less coils)	2" high x 2-5/16" wide x 5-7/8" long.
	(5.08 cm x 5.87 cm x 14.92 cm)
Net Weight (Meter, case, and coils)	2 lbs. (.746 kg.)

It comes with a sturdy plastic case that stores the meter and coils. Each coil covers a specific frequency range and is color coded. The user can make their own coils if desired. Illustration 19 shows the unit with the case and coils.

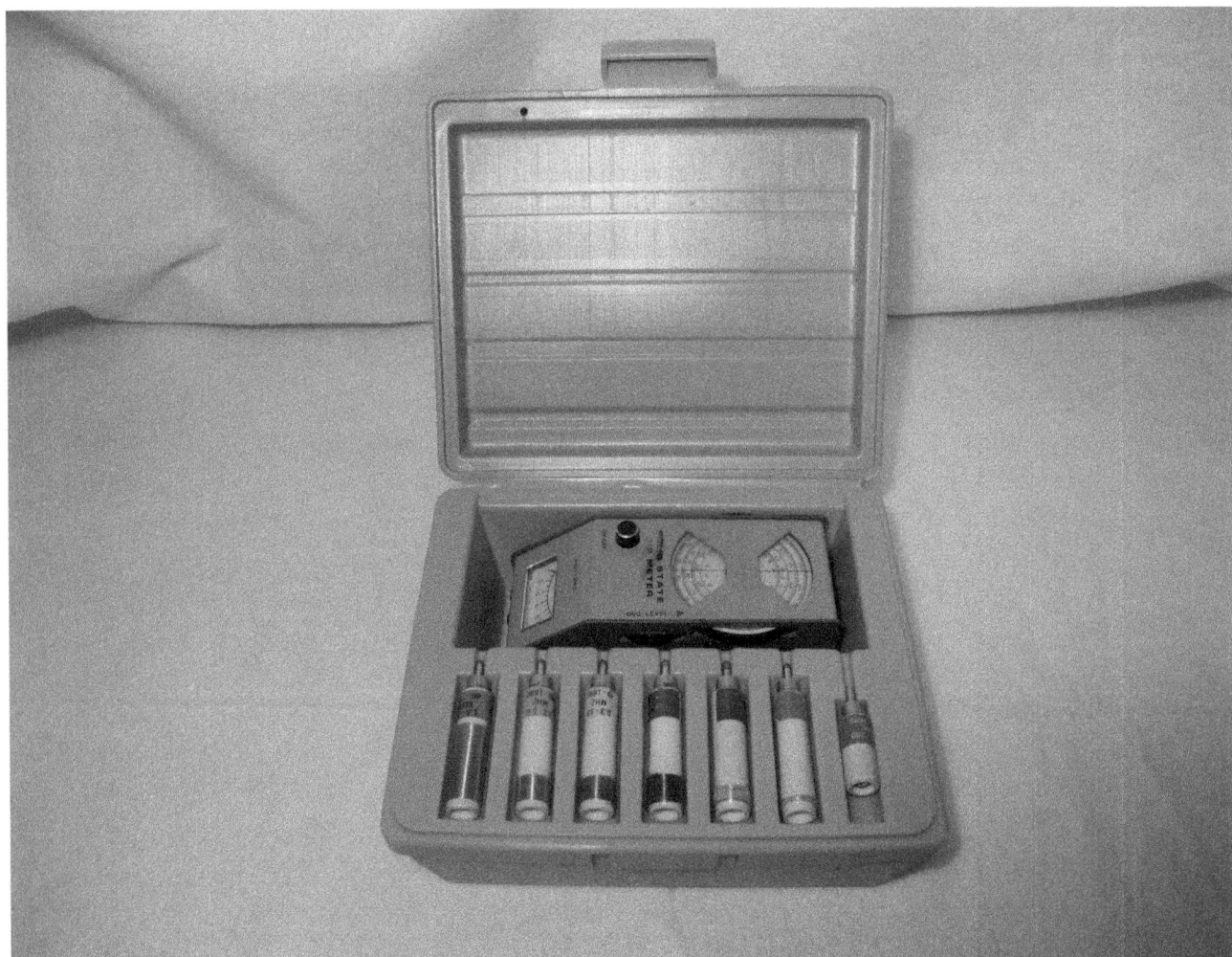

Illustration 19: HD-1250 Front View

It is battery powered and can be held and operated in one hand. On top is the power switch and on the side is the level control which adjusts the output level. The tuning dial adjusts the frequency which can be read on the dial scale. A headphone jack is on the front, although it is rarely used.

The dip meter can operate in two modes: injection mode and absorption mode. An example of injection mode would be measuring the resonant frequency of an LC circuit. The dip meter provides or injects the signal. Given a tuned circuit such as a coil and capacitor in parallel as found in radio circuits, we could couple the meter by placing it near the tuning circuit coil and adjust the frequency until the meter level dips. The dip indicates the resonant frequency of the circuit. The test would be done with the circuit under test powered off.

In absorption mode we can measure the frequency of a signal source. In this mode the dip meter receives or absorbs the signal from an external circuit. In a circuit such as an oscillator we could couple the dip meter by bringing it close to a coil in the circuit when it was in operation and reduce the dip meter level to almost zero. Adjusting the dip meter's tuning dial we should see a peak on the meter when we reach the oscillator frequency.

Another example application is as a field strength meter. Imagine I'm using a two meter band handheld ham radio transceiver to generate a signal. I would use the highest frequency coil, 100 to 250 MHz. Holding the dip meter near the handheld while it is transmitting, and adjusting the tuning until I get a peak, I can read the approximate frequency of the transmitter. Moving around I can measure the field strength, getting information about the antenna. The manual describes more applications for the unit, covering 11 different areas from receiver adjustment to antennas.

The HD-1250 is solid-state, using one bipolar and two FET transistors. Most earlier Heathkit models used vacuum tubes and required AC power. A notable exception was the HM-10/HM-10A Tunnel Dipper which used the relatively little known semiconductor tunnel diode. A view inside the HD-1250 is shown in Illustration 20. While it uses relatively simple circuitry, it is quite compact, being built on two printed circuit boards.

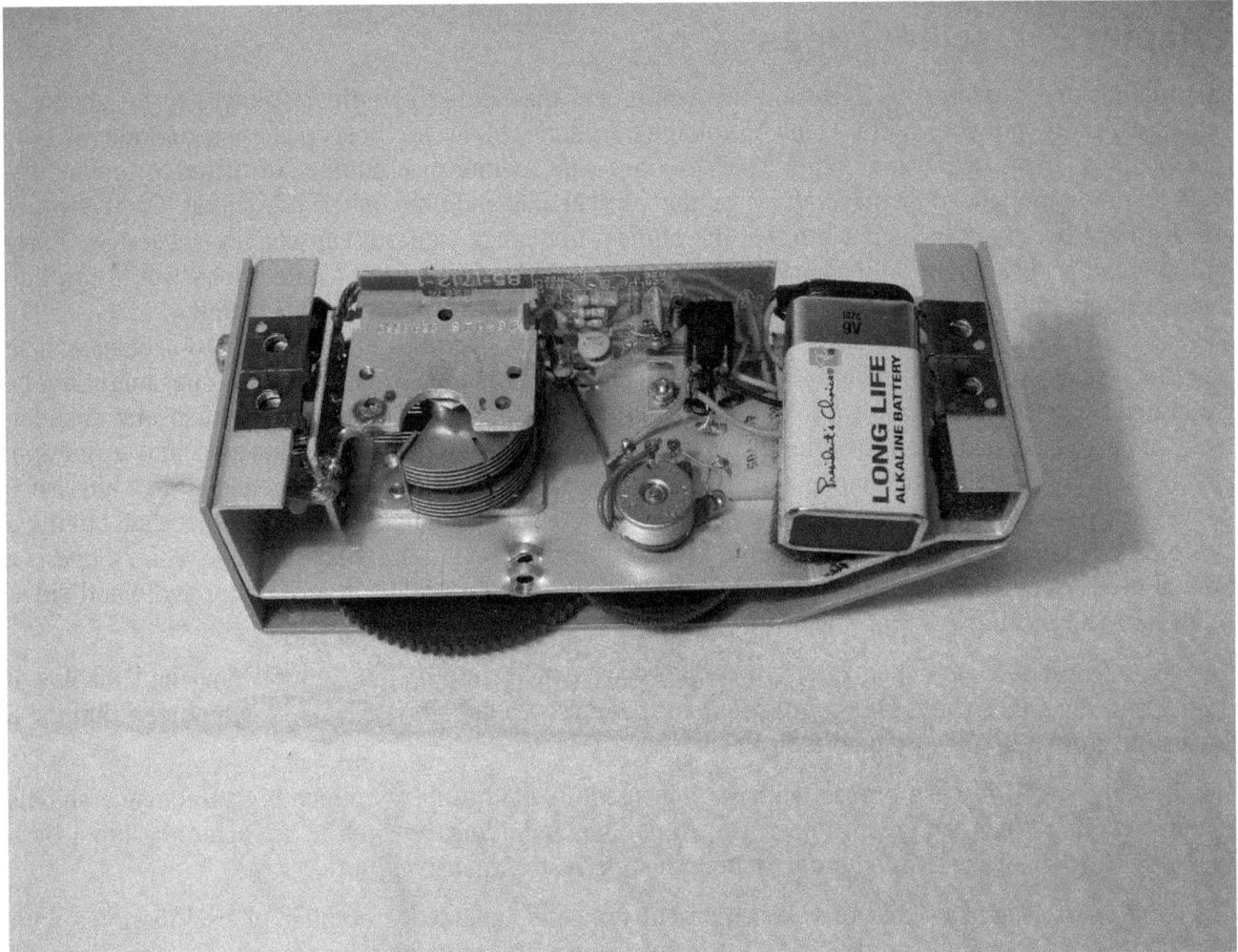

Illustration 20: HD-1250 Inside View

Ideally the unit should be calibrated using a signal of known frequency such as a radio receiver tuned to a shortwave time and frequency standard station such as WWV.

This was another eBay purchase which arrived in good condition and came with the case and all coils. I found a copy of the manual on the Internet. I built an extension probe following the directions in the manual. Because I have a wide assortment of test equipment, I don't tend to use the dip meter a lot, but I have used it as a signal source for testing radio receivers, for testing tuned circuits, and to "neutralize" the final amplifier in my Heathkit DX-60B ham radio transmitter.

Chapter 7: Oscilloscopes

An oscilloscope is a type of electronic test equipment that allows electrical voltages to be observed visually, usually in the form of a graph of voltage over time. Many oscilloscopes use a cathode ray tube (CRT), much like a television. While oscilloscopes can operate in a number of different modes, the most common mode is to display voltage on the vertical axis and time on the horizontal. Oscilloscopes are distinguished by their bandwidth, i.e. the highest frequency signal it can display accurately. Early and low end scopes might only support a bandwidth of audio frequencies while others would run into the megahertz. A higher bandwidth specification generally means a more expensive unit.

The vertical axis typically supports different ranges in steps of voltages in a 1-2-5-10 sequence. The horizontal scale supports different frequencies in a similar manner. Low cost oscilloscopes might not be calibrated in the vertical and/or horizontal axes. Some oscilloscopes support more than one channel, most often two. A dedicated *electronic switch* can turn a single channel oscilloscope into a multiple channel unit by accepting two input signals and outputting them to a scope as one signal that alternates between the two channels. While not as flexible as a true dual channel scope, it it cheaper than buying a new scope to replace a single channel unit. Heathkit offered several of these units. A *vectorscope* is a type of oscilloscope that displays two input signals as an X-Y plot and is often used for audio and video (television) applications.

The above describes so-called *analog oscilloscopes* which amplify the input signal and display it directly on the CRT. There are *digital oscilloscopes*, also called *storage scopes*, which can digitize a signal and store it in memory before displaying it.

Oscilloscopes typically use probes which are designed to faithfully reproduce high frequency signals without loading down the circuit under test, sometimes including a switch to attenuate the input by a factor of 10 when desired. Good probes tend to be relatively expensive.

Over the years Heathkit's line of test equipment typically included a number of oscilloscopes with various price ranges and features. In fact, the company's first electronic kit was the O-1 oscilloscope introduced in 1947. It sold for $49.50, a price that they achieved by offering it as a kit and using some war surplus parts. Illustration 21 shows an early Heathkit oscilloscope kit as well as an electronic switch (that cost almost as much as the scope).

Most of Heathkit's oscilloscopes were offered both as kits and as factory assembled and calibrated instruments. A typical scope is the IO-4205, a dual channel unit with a bandwidth of 5 MHz. While not state of the art today, solid state scopes like the IO-4205 are perfectly suitable for work at lower frequencies, just not high speed digital circuits that operate beyond the scope's bandwidth.

Calibrating an oscilloscope made as a kit typically required either an accurate square wave generator or a dedicated oscilloscope calibrator offered by Heathkit as an additional item. Heathkit also offered scope probes of varying quality and price range.

Digital oscilloscopes are a relatively recent development, and I am only aware of one unit from Heathkit, the ID-4850, which was an external box that would work in conjunction with an analog scope. Table 13 lists the different models of Heathkit oscilloscopes and related accessories.

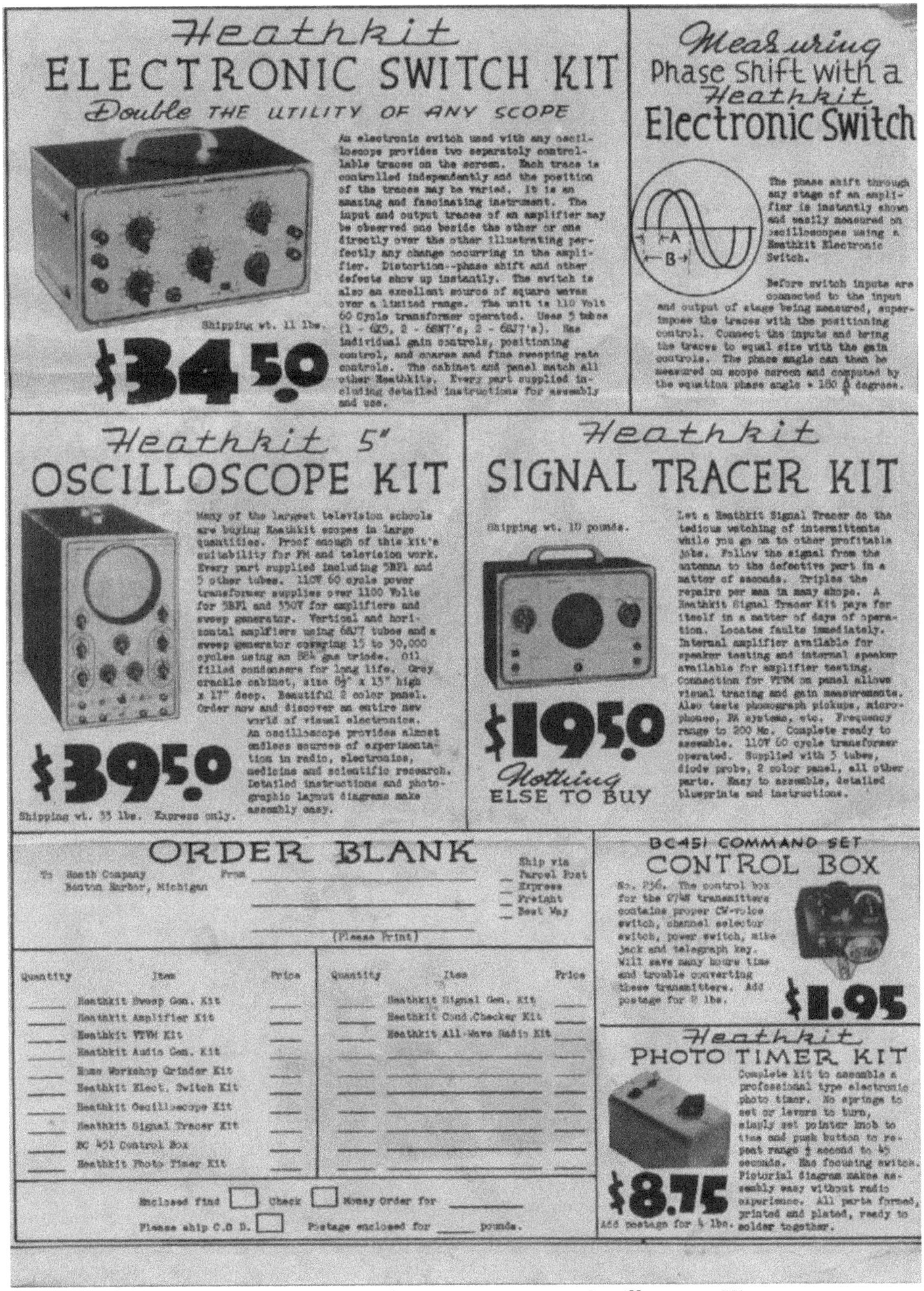

Illustration 21: An Early Heathkit Flyer Featuring an Oscilloscope Kit

Table 13: Heathkit Oscilloscopes and Related Items

Model	Description	First Year	Last Year	Comments
342	Probe - Low Capacitance			X1, Oscilloscope
309-C	Probe - RF			For use with VTVM, scope, signal tracer
337-B	Probe - RF Demodulator			For use with VTVM, scope, signal tracer
337-C	Probe - RF Demodulator			For oscilloscope for signal tracer
EF-2	Oscilloscope Course			Includes test chassis. EF-2-3 includes IO-21. EF-2-5 includes IO-18
EU-70	Oscilloscope	1970	1972	15 MHz, dual trace, solid-state, assembled
EUW-25	Oscilloscope	1963	1970	3", 400 kHz
EV-3	Oscilloscope	1964	1970	IMPScope biological EKG type
EVW-3	Oscilloscope	1968		Assembled version of EV-3
ID-22	Electronic Switch	1964	1970	Converts scope to dual trace
ID-101	Electronic Switch	1971	1976	Converts scope to dual trace
ID-4101	Electronic Switch	1977	1981	Converts scope to dual trace
ID-4850	Digital Memory Oscilloscope	1989	1992	Digital memory box for scopes
IG-4244	Oscilloscope Calibrator	1983	1992	For calibrating oscilloscopes
IG-4505	Oscilloscope Calibrator	1975	1990	For calibrating oscilloscopes
IO-10	Oscilloscope	1960	1967	3", 200 kHz, recurrent sweep
IO-12	Oscilloscope	1962	1968	5", 4 MHz
IO-14	Oscilloscope	1966	1971	5", 8 MHz
IO-17	Oscilloscope	1968	1973	3", 5 MHz
IO-18	Oscilloscope	1968	1970	5", 5 MHz
IO-21	Oscilloscope	1961	1972	3", 200 kHz
IO-30	Oscilloscope	1960	1962	5", 5 MHz
IO-101	Vectorscope/Color Generator	1970	1977	3" vectorscope and color bar/pattern generator
IO-102	Oscilloscope	1971	1975	5", 5 MHz
IO-103	Oscilloscope	1972	1974	5", 10 MHz
IO-104	Oscilloscope	1973	1975	5", 15 MHz
IO-105	Oscilloscope	1971	1974	5", 15 MHz, dual trace
IO-1128	Oscilloscope	1971	1974	3", vector monitor
IO-3220	Oscilloscope	1982	1984	5", 20MHz, dual trace, battery powered
IO-4101	Oscilloscope	1977	1980	Vectorscope like IO-101
IO-4105	Oscilloscope	1979	1987	5", 5 MHz
IO-4205	Oscilloscope	1979	1987	5", 5 MHz, dual trace
IO-4210	Oscilloscope	1989	1990	5", 10 MHz, dual trace
IO-4225	Oscilloscope	1989	1990	5", 25 MHz, dual trace
IO-4235	Oscilloscope	1979	1984	5", 35 MHz, dual trace, delayed sweep
IO-4360	Oscilloscope	1984	1984	5", 60 MHz, triple trace
IO-4510	Oscilloscope	1974	1980	5", 15 MHz, dual trace
IO-4530	Oscilloscope	1975	1977	5", 10 MHz, TV Service

Model	Description	First Year	Last Year	Comments
IO-4540	Oscilloscope	1975	1976	5", 5 MHz, hobby/service
IO-4541	Oscilloscope	1977	1979	5", 5 MHz, special TV triggering
IO-4550	Oscilloscope	1976	1984	5", 10 MHz, dual trace
IO-4555	Oscilloscope	1978	1979	5", 10 MHz
IO-4560	Oscilloscope	1975	1979	5", 5 MHz, auto triggered sweep
IOA-3220-1	Probes - Oscilloscope			Two PKW-105 probes with pouch
IOA-4200	Oscilloscope Timebase Module	1984	1986	Digital display for IO-4360 and IO-4225
IOA-4510-1	Oscilloscope Calibrator			For IO-4510 or IO-4530
IOW-18S	Oscilloscope			Berkeley Physics Laboratory, 5" laboratory
O-1	Oscilloscope	1947	1948	5"
O-2	Oscilloscope	1948		5"
O-3	Oscilloscope	1948	1948	5", 150 kHz
O-4	Oscilloscope	1949	1949	5", 2 MHz
O-5	Oscilloscope	1950	1950	5", 2.2 MHz
O-6	Oscilloscope	1950	1951	5", 200 kHz
O-7	Oscilloscope	1951	1951	5", 250 kHz
O-8	Oscilloscope	1951	1953	5", 2 MHz
O-9	Oscilloscope	1951	1954	5", 3 MHz
O-10	Oscilloscope	1955	1956	5", 400 kHz, PC board
O-11	Oscilloscope	1957	1957	5", 5 MHz
O-12	Oscilloscope	1958	1990	5", 5 MHz
OL-1	Oscilloscope	1955	1956	5", 5 MHz
OM-1	Oscilloscope	1955	1956	5", 5 MHz
OM-2	Oscilloscope	1957	1957	5", First Heathkit product
OM-3	Oscilloscope	1958	1960	5", 1.2 MHz
OP-1	Oscilloscope	1958	1962	2.2 MHz
OR-1	Oscilloscope	1959	1962	5", 200 kHz
PKW-2	Probe - Oscilloscope			X1 X10 25 MHz
PKW-101	Probe - Oscilloscope			X10 60 MHz
PKW-104	Probe - Oscilloscope			X1 17 MHz
PKW-105	Probe - Oscilloscope			X1 15 MHz, X10 80 MHz
S-1	Electronic Switch	1950	1950	Dual trace for scope
S-2	Electronic Switch	1950	1955	Dual trace for scope
S-3	Electronic Switch	1956	1962	Dual trace for scope
SDS-5000	Oscilloscope	1988	1988	Computer-based
SO-29	Oscilloscope	1972	1972	Biological, high gain DC
SO-3220	Oscilloscope			Assembled version of IO-3220
SO-4105	Oscilloscope			Assembled version of IO-4105
SO-4205	Oscilloscope			Assembled version of IO-4205
SO-4221	Oscilloscope		1987	5", 20 MHz, dual trace

Model	Description	First Year	Last Year	Comments
SO-4226	Oscilloscope	1987	1989	5", 25 MHz, dual trace
SO-4510	Oscilloscope			Assembled version of IO-4510
SO-4521	Oscilloscope	1987	1989	5", 50 MHz, dual trace
SO-4530	Oscilloscope			Assembled version of IO-4530
SO-4540	Oscilloscope			Assembled version of IO-4540
SO-4550	Oscilloscope			Assembled version of IO-4550
SO-4552	Oscilloscope	1989	1992	5", 25 MHz
SO-4554	Oscilloscope	1989	1992	5", 40 MHz
SU-511-50	50 Ω Termination			For counters, oscilloscopes, DC to 1 GHz
VC-1	Oscilloscope Calibrator	1951	1951	For calibrating oscilloscopes
VC-2	Oscilloscope Calibrator	1953	1956	For calibrating oscilloscopes
VC-3	Oscilloscope Calibrator	1957	1962	For calibrating oscilloscopes

Let's review the inputs and controls commonly found on oscilloscopes. Illustration 22 shows the front panel of the Heathkit IO-4205, a model that is typical of most oscilloscopes and one that we will look at in depth later in this chapter. Here is a brief description of each of the controls and inputs.

INTENSITY - Clockwise rotation increases the brightness of the display. Adjusted as necessary for room lighting conditions. Refocusing may be necessary when the brightness is changed. CAUTION: Do not allow a bright spot to remain on the screen as it could damage the CRT.

FOCUS - Varies the shape and size of the beam striking the face of the CRT. Adjusted for the sharpest display.

TRIG LEVEL - Adjusts the trigger circuit so the sweep can be started at any position on the input signal waveform.

HORIZ POS - Positions the trace horizontally on the screen.

Y1 POS - Positions the channel Y1 trace vertically on the screen.

Y1 INPUT - This is the input connector for channel Y1.

AC-GND-DC (Input switch) - This switch is provided for each input channel. In the **AC** position, this switch blocks the DC level of the input signal so that only the AC component is displayed. In the **GND** position, the input is disconnected and the vertical amplifier input is grounded. This position is used when you wish to set the baseline (trace) at a desired position without disconnecting the input signal. In the **DC** position, both DC and AC components of the input signal are displayed.

VOLTS/CM - This switch is provided for each input channel. Each position of the attenuator switch is marked for the number of volts (peak-to-peak) required to produce a pattern one centimeter high on the graticule.

VOLTS/CM VARIABLE - This control is provided for each input channel. It is normally operated in its fully clockwise (**CAL**) position where the **VOLTS/CM** switch positions are calibrated. Vertical gain decreases as the control is turned counterclockwise, permitting the vertical trace size to be adjusted.

However, the display is then uncalibrated.

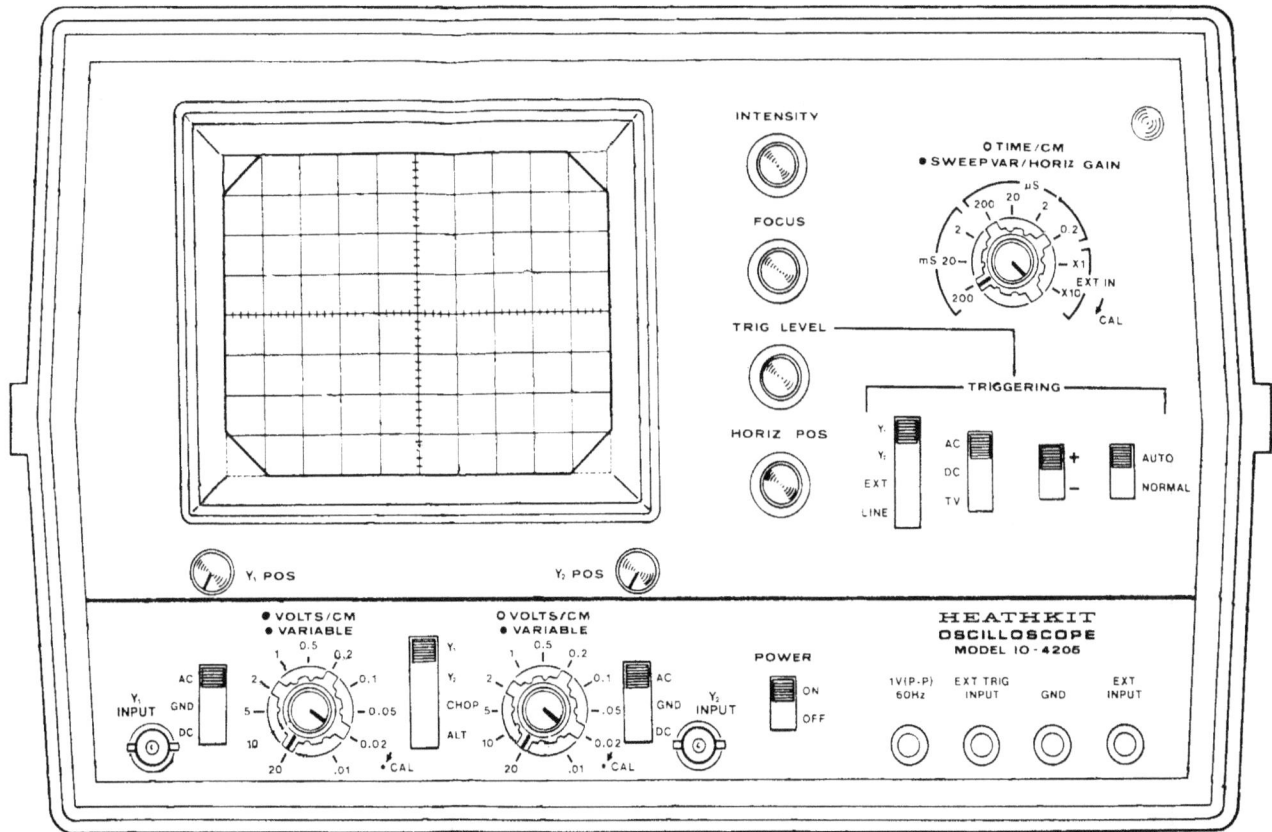

Illustration 22: IO-4205 Front Panel Controls

VERTICAL MODE switch (**Y1-Y2-CHOP-ALT**) - Displays either channel Y1, Y2, or both when in the **CHOP** or **ALT** position. In the **CHOP** position, the two inputs are sampled at approximately a 100 kHz rate and displayed. In the **ALT** position, the horizontal sweep alternates between inputs. A complete sweep of one input is displayed and then a complete sweep of the other input signal is displayed.

Y2 POS - Positions the channel Y2 trace vertically on the screen.

Y2 INPUT - This is the input connector for channel Y2.

POWER - Turns the oscilloscope on and off.

POWER LAMP - Glows then AC power is turned on.

TIME/CM - The time required for the beam to sweep one centimeter is determined by the **TIME/CM** switch when the **SWEEP VAR/HORIZ GAIN** control is fully clockwise (**CAL**). Counterclockwise rotation decreases the sweep speed. In the **EXT IN** positions, the signal at the **EXT INPUT** connector is coupled to the horizontal amplifier; the **SWEEP VAR** control then adjusts the horizontal gain (**HORIZ GAIN**).

TRIGGER SOURCE switch (**Y1, Y2, EXT. LINE**) - Connects the trigger circuits to the Y1 trigger signal, the Y2 trigger signal, an external trigger signal, or a 60 Hz internal signal.

TRIGGER COUPLING switch (**AC-DC-TV**) - The **DC** position couples the trigger signals directly to the trigger circuits. This allows the sweep to be triggered from DC level changes or very low frequency AC signals. In the **AC** position, the DC component of the trigger signal is blocked so that only the AC component of the signal reaches the trigger circuits. The **TV** position cuts off unwanted high frequency signals so you can lock onto TV vertical frame signals.

SLOPE switch (+/-) - The sweep can be started on either a positive or negative slope, depending on the position of the +/- switch.

TRIGGER MODE switch (**AUTO-NORMAL**) - In the **AUTO** position, a base line will always be present in the absence of a trigger signal. In the **NORMAL** position, the base line is not automatically generated.

1V(P-P) 60Hz - Provides a 0.94 volt, peak-to-peak, 60 Hz sine wave signal for testing or calibrating. Many oscilloscopes label this as **CAL** and output signals of different amplitudes and frequencies depending on the oscilloscope (a 1 kHz square wave is common).

EXT TRIG INPUT - An external signal can be applied through this connector to trigger the sweep circuits when the **TRIGGERING** switch is in the **EXT** position.

GND - Provides a connection to circuit ground.

EXT INPUT - Allows you to apply an external X-input signal. (A positive signal moves the trace to the right.)

In-Depth: The IO-4205 Dual Channel Oscilloscope

Let's look at a typical model of 'scope offered by Heathkit, the IO-4205 Dual Trace Oscilloscope.

The IO-4205 was a dual channel oscilloscope offered by Heathkit from 1979 to 1987. My 1982 Canadian Heathkit catalog lists it at a price of $549.95 as a kit and $999.95 fully assembled (the assembled version was the SO-4205). I have seen another undated US catalog that listed it at $329.95 for the kit and $480.00 assembled.

This was a mid-range scope. The IO-4105 was a lower priced unit that was essentially the same but with a single trace, for about $150 less. At the high end, my 1982 catalog lists the 35 MHz IO-4235 scope for $1799.95 as a kit.

In addition to the scope, you would need one or two probes that sold for $59.95 each and if building it as a kit, you would typically need to purchase the oscilloscope calibrator kit at $119.95 (these prices are from 1982).

The single most important specification of an oscilloscope is its bandwidth, the range of frequencies over which it can accurately display signals. The IO-4205 is a DC to 5 MHz scope. At the time, this was pretty reasonable and suitable for most analog and digital circuits. This is a dual channel scope that provided two independent channels, although they share the same horizontal timebase circuit.

The display is 8 by 10 centimeters. The vertical sensitivity went as low as 10 millivolts per centimeter with 11 calibrated voltage ranges in a 1-2-5 sequence. The horizontal sweep rate went from 200 ms to

0.2 microseconds per division in 7 ranges.

It provides channel 1, channel 2, external, and line trigger sources with AC, DC, or TV trigger modes, plus or minus slope, automatic or normal trigger.

The advantage of a dual channel scope is that you can display two waveforms at once. It can also operate in an X-Y mode where one channel drives the vertical and one the horizontal. A Z axis input, which controls intensity, is not supported by this scope.

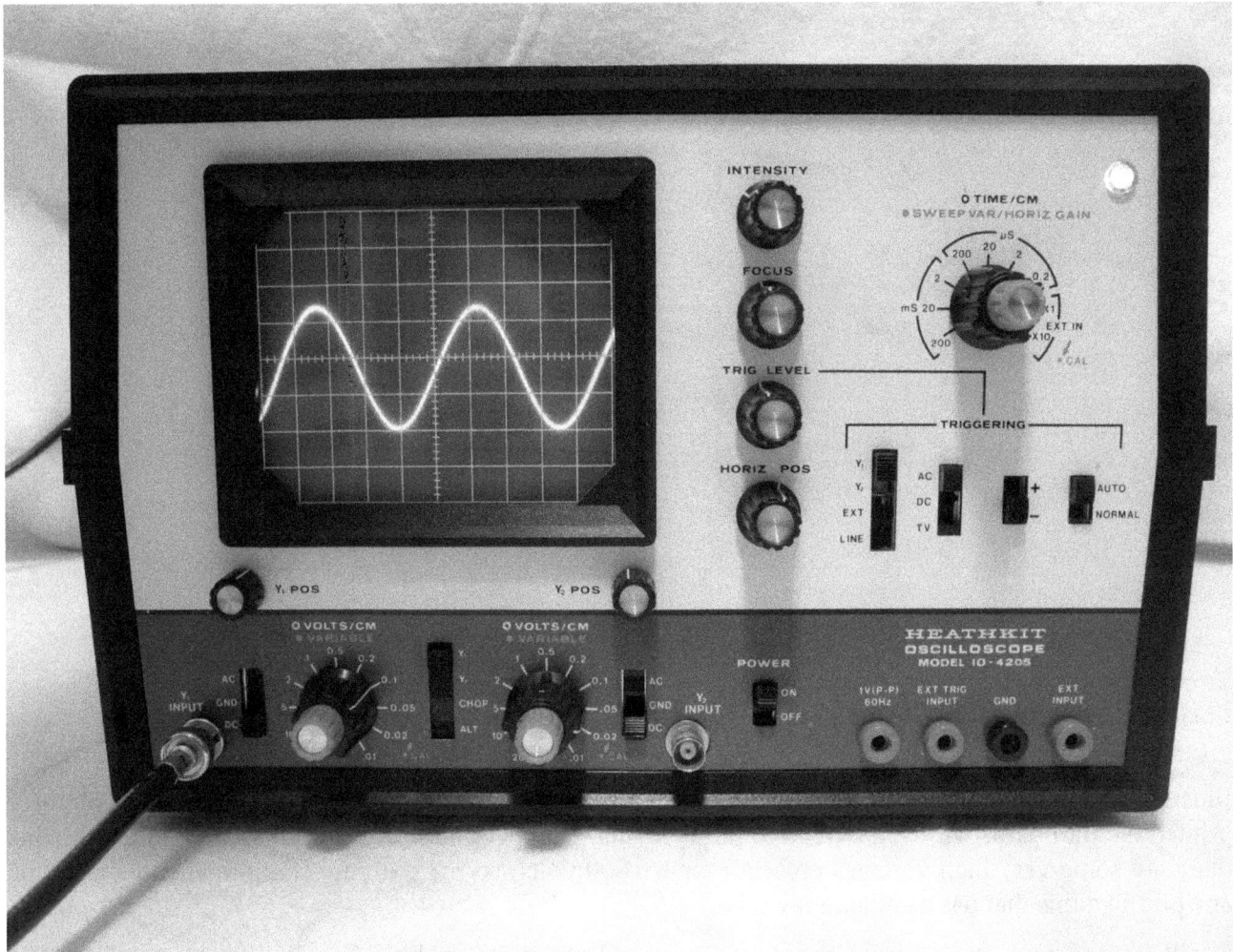

Illustration 23: IO-4205 Front View

Illustration 23 is a front view of the scope displaying a one megahertz sine wave from a signal generator. The amplitude of the input signal is about 3.5 volts peak to peak and the period is about one microsecond, corresponding to 1 MHz. If we increase the frequency of the input signal the displayed amplitude drops off as we exceed the bandwidth of the scope. It was rated as within ±3 dB to 5 MHz. My testing shows that it drops off quite slowly and higher frequencies can still be observed, however with reduced amplitude.

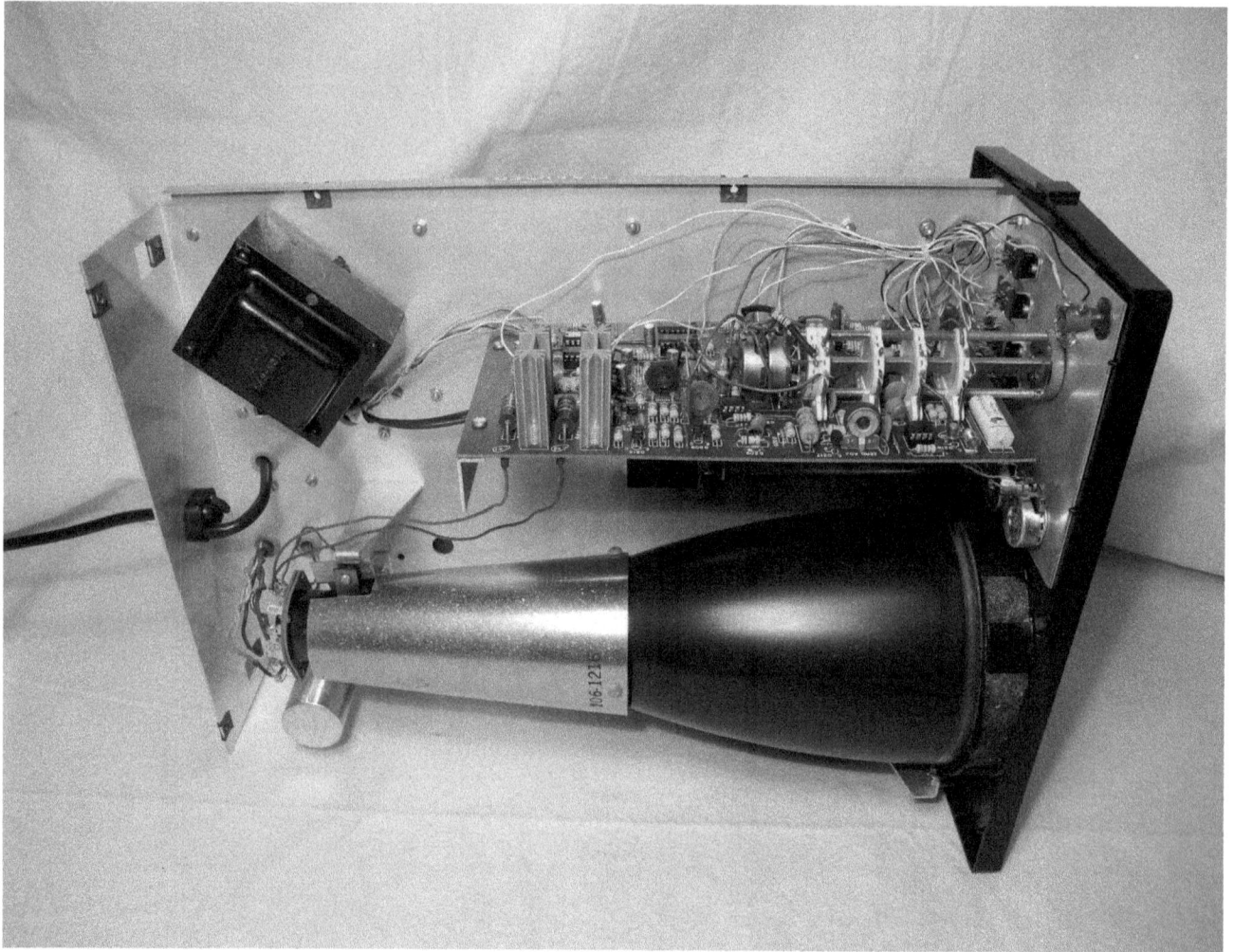

Illustration 24: IO-4205 Inside View

Illustration 24 shows a view inside the unit. It is all solid state except for the CRT. Circuitry is on three printed circuit boards with some point to point wiring. Whoever built this unit did a nice job. Note that there are some very high voltages present (about 1600 volts) so use extreme caution when working on any unit like this that has a cathode ray tube.

Oscilloscopes were always quite expensive items and not many hobbyists could afford a new one. The IO-4205 was quite a good value for an oscilloscope. There were less expensive scopes on the market, but they tended to have lower bandwidth (sometimes only audio frequencies), not be calibrated in the vertical or horizontal, and sometimes have a simple recurrent rather than triggered sweep. The IO-4105, the single channel version, was even more affordable.

I purchased this unit in October 2011 from a local radio amateur along with a non-Heathkit digital scope and a Heathkit frequency counter. I cleaned it up a little but not much else was needed. I bought some low cost modern scope probes to use with it. The calibration was a bit off. I was able to find a schematic on the Internet but not a full manual, so I purchased a manual from one of the commercial sources. I performed a complete checkout and calibration of the unit using the procedure in the manual.

The unit works well but to be honest I purchased it more as a collector's item rather than to use on a daily basis. My regular scope is a more modern BK Precision 30 MHz dual channel scope.

When I was about 16 years old I dreamed of owning a decent oscilloscope and this was one of the models that I looked at. At around $500 for even the single channel version, it was too expensive to justify for a student who made about $4 an hour at a part-time job. Finally getting to own one years later is a real kick, even if it is no longer state of the art, and it is still perfectly useful for working on older analog radios and test equipment.

Chapter 8: Power Supplies

I consider power supplies as a type of test equipment because they are frequently found on a test bench and used for testing. Heathkit had quite a complete line of power supplies over the years. Much of the tube-based Heathkit amateur radio equipment required separate, specialized, power supplies. I don't include these within the scope of this book.

Power supplies are mainly characterized by the voltage levels that they output. Roughly speaking they can be grouped into low voltage (about 30 volts and below) and high voltage (above 30 volts). Vacuum tube circuits tend to require high voltages, anywhere from about 150 VDC to 500 VDC or more while solid state circuits generally operate on low voltages, often 12 volts or 5 volts for digital circuits. Tubes also typically require either 6.3 VAC or 12.6 VAC to power the tube filaments (heaters).

The second important characteristic is the maximum current that a supply can produce, in amps or milliamps. Generally speaking a greater current capacity implies a larger and more expensive power supply.

A given output may be fixed in voltage, or variable. A variable output provides more flexibility while some voltages, such as 5V and 12V are so commonly used that a fixed output supply for these is useful.

As well as voltage, some variable supplies provide a facility for controlling the maximum output current; these are sometimes called constant current supplies when used in this mode. If a current control is not provided there is usually some way of limiting the current drawn, either limiting it to a fixed maximum and/or a fuse or circuit breaker.

Supplies may provide meters, either analog or digital, to display the output voltage and/or current.

The specification for *load regulation* refers to how closely the supply maintains the output voltage under changing loads, usually specified as a percentage. *Line regulation* refers how closely the output remains constant when the input line voltage changes. DC power supplies may also be rated for the amount of AC ripple and noise in the output.

Some power supplies have *floating outputs*, which means that they have no voltage reference to each other or to ground and can therefore be connected together to combine outputs or connected to a device without using a common ground.

Some old equipment has a chassis that is at a high voltage. This type of equipment can be dangerous to operate because touching the chassis while grounded will cause a shock. By powering the equipment from an *isolation transformer*, there is no reference to ground and therefore a reduced risk of shock.

When powering up older vacuum tube equipment it is often prudent to start with a low voltage and gradually increase it while examining the equipment for a failure. This can be done using a variable autotransformer. Sometimes both a variable autotransformer and isolation transformer are incorporated into a power supply; Heathkit made several models like this.

For solid-state electronics work I would recommend as a good supply one with fixed +5V, +12V, and -12V outputs and a variable 0 to 20 V output with voltage and current meters. I typically have one of these permanently on my bench. For working on tube-based equipment I would recommend a high voltage power supply that also provides 6.3 and 12.6 VAC (for tube heaters). However, most tube-based equipment incorporates its own power supply, so it is really most useful if you are constructing your own equipment. A variable autotransformer and isolation transformer are useful for working on tube equipment such as old radios, either separate units or a power supply that incorporates both.

Table 14 lists all of the power supply models that I am aware of with a brief summary of their specifications. When output voltages are listed they are DC voltages unless indicated otherwise.

Illustration 25 shows the IP-2700 series of power supplies that were introduced in 1975.

Table 14: Heathkit Power Supplies

Model	Description	First Year	Last Year	Comments
ES-1	Power Supply	1956	1956	For use with ES-400 computer
ES-100	Power Supply	1956	1956	For use with ES-400 computer
ETI-7030	Power Supply	1990	1990	Bench type, triple output
EU-40	Power Supply	1969	1970	0 to 300V @ 20mA, 6.3 VAC @ 1A
EU-40A	Power Supply			50 to 300V @20mA, 6.3VAC @1A
EU-41	Power Supply	1969	1970	0 to 15V @ 750mA
EU-41A	Power Supply			0 to 15V @750mA
EUW-15	Power Supply	1964	1970	200 to 350 VDC @ 100mA, 6.3VAC @ 3A
EUW-17	Power Supply	1967	1970	0 to 25V @ 200mA
IP-10	Isolation Transformer	1961	1962	90 to 130 VAC, 300W, meter
IP-12	Power Supply	1962	1975	Battery Eliminator. 6V unfiltered @10A, filtered @5A or 12V unfiltered @ 5A, filtered @5A
IP-17	Power Supply	1968	1977	0 to 40V @100mA, 0 to -100V @1mA, 6.3VAC @4A, 12.6VAC @2A
IP-18	Power Supply	1968	1977	1 to 15V @ 500mA
IP-20	Power Supply	1962	1967	0 to 50V @ 1.5A
IP-22	Isolation Transformer	1963	1964	90 to 130VAC, 300W, meter
IP-27	Power Supply	1968	1975	0.5 to 50V @1.5A, meter
IP-28	Power Supply	1969	1976	1 to 10V or 1 to 30V @1A, meter
IP-32	Power Supply	1962	1967	0 to 100V @ 1mA, 0 to 400V @ 100mA
IP-2670	Power Supply	1975		7.5 to 15V, meters
IP-2700	Power Supply	1975	1976	0 to 60V @ 1.5A, analog meters
IP-2701	Power Supply	1975	1977	0 to 60V @ 1.5A, digital meters
IP-2710	Power Supply	1975	1981	0 to 30V @ 3A, analog meters
IP-2711	Power Supply	1975	1981	0 to 30V @ 3A, digital meters
IP-2715	Power Supply	1975	1983	Battery eliminator. 9-15V@12A, meters
IP-2717	Power Supply	1977	1982	0 to 400 V @ 125mA, 0 to 100V @ 1mA, 6.3VAC, 12.6VAC
IP-2717A	Power Supply	1982	1991	0 to 400V @ 100mA, 0 to 100V@1mA. 6.3VAC@4A, 12.6VAC @ 2A, meters
IP-2718	Power Supply	1976	1992	5V @1.5A, 2 X 0 to 20V @0.5A, meter
IP-2720	Power Supply	1975	1977	0 to 15V @ 5A, analog meters
IP-2721	Power Supply	1975	1976	0 to 15V @ 5A, digital meters
IP-2728	Power Supply	1977	1990	1 to 15V @ 500mA

Model	Description	First Year	Last Year	Comments
IP-2730	Power Supply	1975	1977	0 to 7.5V @ 10A, analog meters
IP-2731	Power Supply	1975	1977	0 to 7.5V @ 10A, digital meters
IP-2760	Power Supply	1984	1987	Battery eliminator - restyled IP-2715
IP-5220	Power Supply	1975	1983	0 to 140VAC with isolation
IPA-5280-1	Power Supply	1977	1989	Power supply for IT-5280 series instruments
IPW-17	Power Supply			Berkeley Physics Laboratory
IPW-27	Power Supply	1968	1975	Assembled version of IP-27
IS-1	Isolation Transformer			Two voltage settings, meter
IT-1	Isolation Transformer	1953	1960	Two voltage settings, meter
PS-1	Power Supply	1950	1951	50 to 300V
PS-2	Power Supply	1951	1954	160 to 450V, 6.3 VAC @ 4A
PS-3	Power Supply	1957		500V @ 200mA, 6.3VAC
PS-4	Power Supply	1961		0 to 400V @ 100 mA, 0 to -100 V @ 1 mA, 6.3VAC @ 4 A, meters
SP-17A	Power Supply			Assembled version of IP-17
SP-18A	Power Supply			Assembled version of IP-18
SP-2700	Power Supply			Assembled version of IP-2700
SP-2701	Power Supply			Assembled version of IP-2701
SP-2710	Power Supply			Assembled version of IP-2710
SP-2711	Power Supply			Assembled version of IP-2711
SP-2717	Power Supply	1981		Assembled version of IP-2717A
SP-2717A	Power Supply			Assembled version of IP-2717A
SP-2718	Power Supply		1992	Assembled version of IP-2718
SP-2720	Power Supply			Assembled version of IP-2720
SP-2721	Power Supply			Assembled version of IP-2721
SP-2730	Power Supply			Assembled version of IP-2730
SP-2731	Power Supply			Assembled version of IP-2731
SP-2762	Power Supply	1989	1991	0 to 30V @ 3A, analog meters

Digital & Analog Supplies

Four DC models — 0.7.5V @ 10A, 0-15V @ 5A, 0-30V @ 3A, 0-60V @ 1.5A. 3½-digit or 3½" meter readout. Features include: Remote programming of voltage and current through rear panel connectors using external sources. Adjustable constant current and adjustable constant voltage modes. Remote voltage sensing capability. Output protection against short-circuit operation and accidentally applied voltages. Load protection against open remote-sensing leads.

Kit IP-2700, 60V analog,
Shpg. wt. 35 lbs. **169.95**

Assembled SP-2700, 60V analog,
Shpg. wt. 35 lbs. **255.00**

Kit IP-2701, 60V digital,
Shpg. wt. 37 lbs. **219.95**

Assembled SP-2701, 60V digital,
Shpg. wt. 37 lbs. **340.00**

Kit IP-2710, 30V analog,
Shpg. wt. 34 lbs. **169.95**

Assembled SP-2710, 30V analog,
Shpg. wt. 34 lbs. **255.00**

Kit IP-2711, 30V digital,
Shpg. wt. 36 lbs. **219.95**

Assembled SP-2711, 30V digital,
Shpg. wt. 36 lbs. **340.00**

Kit IP-2720, 15V analog,
Shpg. wt. 34 lbs. **169.95**

Assembled SP-2720, 15V analog,
Shpg. wt. 32 lbs. **255.00**

Kit IP-2721, 15V digital,
Shpg. wt. 36 lbs. **219.95**

Assembled SP-2721, 15V digital,
Shpg. wt. 34 lbs. **340.00**

Kit IP-2730, 7.5V analog,
Shpg. wt. 36 lbs. **169.95**

Assembled SP-2730, 7.5V analog,
Shpg. wt. 34 lbs. **255.00**

Kit IP-2731, 7.5V digital,
Shpg. wt. 38 lbs. **219.95**

Assembled SP-2731, 7.5V digital,
Shpg. wt. 36 lbs. **340.00**

IP/SP-2700 SERIES SPECIFICATIONS

(NOTE: Specs measured in accordance with NEMA standards (PYI-1972) after 30-minute warmup period.)

Max. Rated Output: Models IP/SP-2700 & 2701; 60V, 1.5A; models IP/SP-2710 & 2711: 30V, 3A; models IP/S-2720 & 2721: 15V, 5A; models IP-2730 & 2731: 7.5V, 10A. **Load Regulation:** Voltage ± 0.05% +1 mV, Current: ±0.10% + 3.5 mA. **Line Regulation:** Voltage: ± 0.05% + 1 mV. Current: ±0.10% + 1 mA. Ripple & Noise: Voltage: 1 mV RMS; 0.03% of rated output, peak-to-peak. **Voltage/Current Readout** (Switchable): Analog: 3½", 100°, meter. Digital: 3½-digit (1999), two-decade, auto-ranging, digital meter. **Readout Accuracy:** Voltage: Analog — ±3% of rated output; Digital — ±0.5% of reading ± 1 count using lab standards, ±1.0% of reading ± 1 count using built-in calibrator. Current: Analog — ±3% of rated output; Digital — ±1.0% of reading ± 4 count using lab standards, ± 1.5% of reading ± 4 count using built-in calibrator. Readout Response Time: (Digital): 2 seconds to within 5 counts. Stability at Output Terminals: Voltage: ± (0.01% + 1 mV)/hr. Current: ± (0.05% + 1 mA)/hr. Stability as Displayed (Digital): Voltage: ± (0.01% + 1 mV + ½ count)/hr. Current: ± (0.01% + 1 mA + ½ count)/hr. Operating Modes: Constant voltage, constant current, auto-series, auto parallel. Programming Mode: Voltage: A—zero to rated output with 0 to 5.0 VDC applied; B—zero to rated output with 0 to 5000 Ω external resistor. Current: Zero to rated output with applied voltage of 1.0 volt/amp (0.1 volt/amp for IP-2731). Frequency Response: DC to 100 Hz, ± 2 dB. Programming Transient Response: 0.1 ms for low current to high current changes; 1.0 ms for high to low. Load resistance less than 10 x (E **Rated**/I **Rated**). Power Requirements: 120/240 ± 10/20 VAC; 60/50 Hz, 2.0/1.0 Amps max. **Overall Dimensions:** 5.5" H x 15" W x 13.5" D.

Illustration 25: IP-2700 Series Power Supplies from 1976 Catalog

In-Depth: The IP-27 Low-Voltage Regulated Power Supply

The IP-27 is a solid-state regulated low-voltage power supply that produces a variable output from 0.5 to 50 volts with adjustable current limiting up to 1.5 amps. It was produced from 1968 to 1975 and offered as kit or factory assembled, the factory assembled version being the IPW-27. My 1971 catalog lists it at a price of $79.95 and the assembled IPW-27 for $125.00

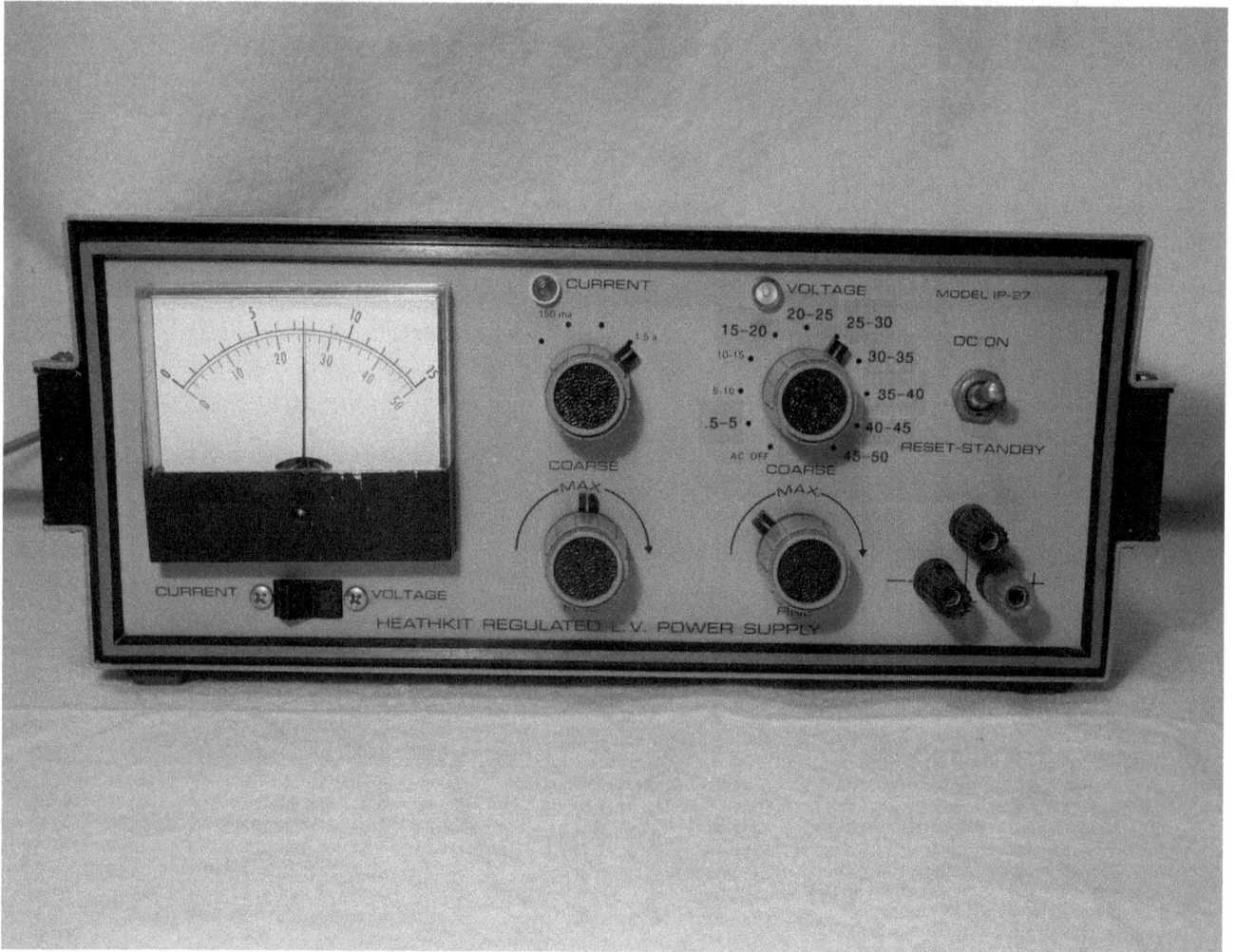

Illustration 26: IP-27 Front View

It was described there as "The Finest Low Voltage Supply Heath Has Ever Offered". The specifications, taken from the Heathkit catalog, are as follows:

SPECIFICATIONS-Input: 105-125 volts or 210-250 volts, 50.60 Hz; 135 watts at full load (50C, 1.5A). **Output:** 0.5-50 volts DC; 1.5 amps max. DC. **Load regulation:** +/- 0.05% mV, can be adjusted for no change. **Line regulation:** For a 10% change in line voltage, less than 0.05% change. **Ripple & noise:** Less than 250 microvolts. **Transient response:** Less than 25 microseconds. **Output impedance:** Less than 0.075 ohms from DC to 10 kHz, less than 0.3 ohms 10 kHz & up. **Overload protection:** Current limiter & relay. **Meter:** 3-1/2", 1 mA, 50 ohm. **Current ranges:** 50 mA, 150 mA, 500 mA, 1.5 A. **Voltage ranges:** 5V, 15V, 50V. **Front panel controls:** On/Off-Coarse voltage switch, Fine voltage adjust, Current range switch, Current limit adjust, Meter rocker switch, Reset Standby toggle switch. **Output terminals:** (3), plus (+), minus (-), chassis ground. **Dimensions:** 5-1/8" H x 13-1/4" W x 9" D.

Illustration 26 shows the front panel. A meter shows output voltage or current depending on the position of the switch below it, with neon lamps indicating the mode.

The Coarse Voltage switch selects the output voltage in one of 10 ranges that each cover a 5 volt range over the full range from 0.5 to 50 volts. The Fine Voltage control adjusts the output voltage continuously within the selected range.

Similarly, output current limiting can be set in one of four ranges using the Coarse Current switch: 50 mA, 150 mA, 500 mA, and 1.5 amps. The Fine Current control adjusts the output current continuously over the range.

Red and black colors on the voltage and current range switch markings correspond to the meter scale to read for that range.

A toggle switch selects DC On or Reset-Standby. This allows you to set the unit for specific output voltage and current and then turn off the output, such as when connecting or disconnecting the supply to a circuit under test.

The output is protected by current limiting as well as by a relay which protects it against overloads or shorts.

The output is floating. There are plus and minus outputs as well as ground. Multiple units can be connected in series for more voltage or in parallel for more current.

Illustration 27 shows a view inside the chassis. All wiring is point to point. It has a large power transformer and several large electrolytic filter capacitors. The design is all solid state but no ICs are used.

The calibration procedure only requires a power resistor to use as a load (and in a pinch the manual says you can use an electric iron). It uses the built-in meter for adjustment. The unit seems to run quite cool even at full output current.

I purchased an IP-27 on eBay in 2013 from a seller who had several units. It appears to have been used by a marine electronics company for lab or manufacturing purposes. According to a sticker on the unit it was last calibrated in 2008. I suspect it was in use there for many years. The unit as received was fully working. The outside of the case was a little worn but it was clean inside. It did not come with a manual but I found a full copy of the manual on the Internet. It produced full output voltage and current and running through the entire calibration procedure I found it was bang on and did not need any adjustment.

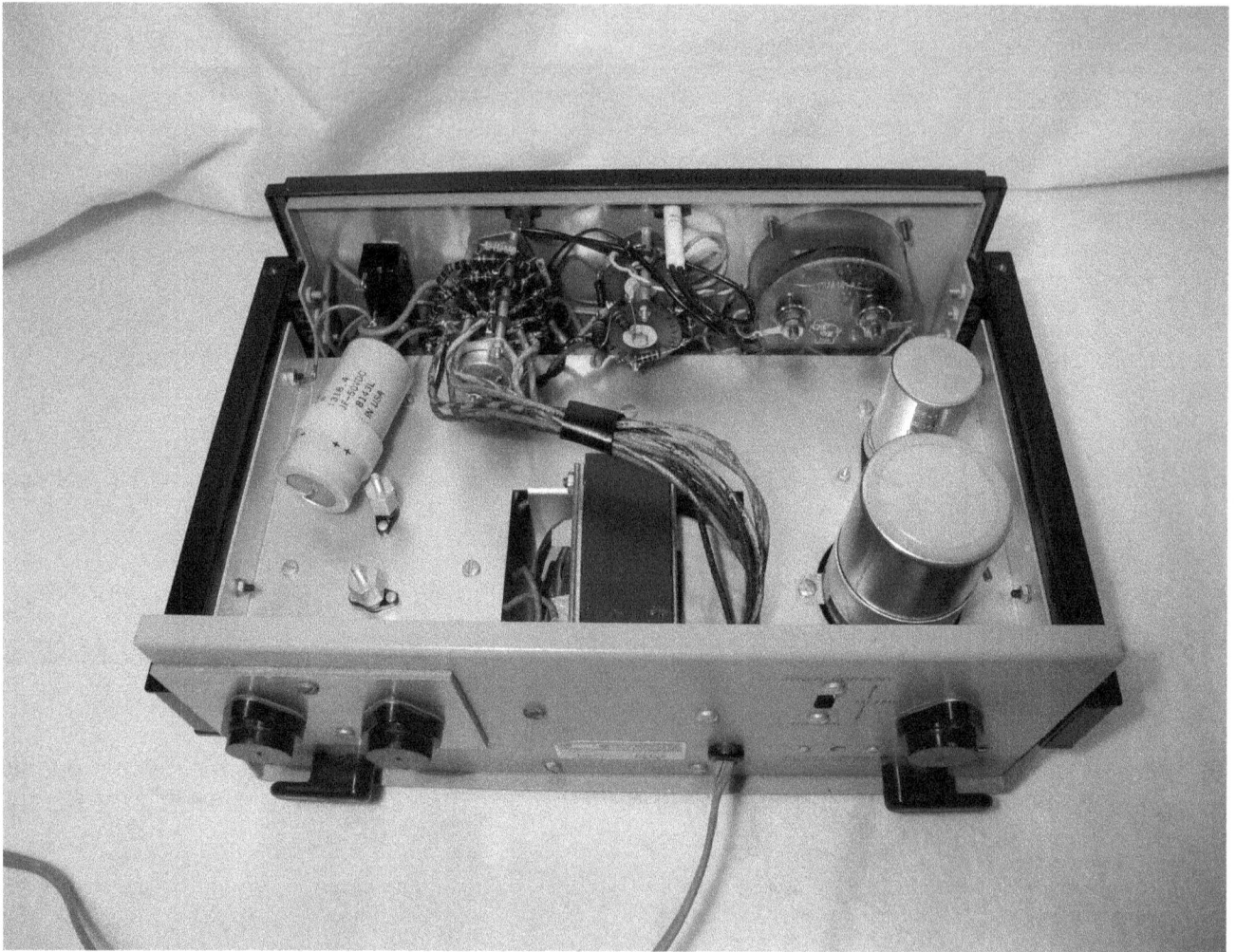

Illustration 27: IP-27 Inside View

The front panel dial markings that were in red ink had faded so I replaced them with labels from a label printer. An Internet search found that Heathkit recommended a modification to prevent possible transistor failure when the voltage switch is changed rapidly. My unit had not had that modification made, so I applied it.

The power supply uses germanium transistors which may be difficult to find replacements for.

In summary, the IP-27 was Heathkit's high end low voltage power supply and undoubtedly many of them were sold over the eight years that the product was offered. Many of them, like mine, were probably in daily use for decades. Having different voltage and current ranges is a nice touch - lower quality supplies tend to use one range for output adjustment which made it hard to adjust accurately.

In 1975, the IP-27 was superseded by the IP-2700 series of power supplies which offered analog or digital meters and models having four different maximum output voltages.

Chapter 9: Signal Generators

Signal generators are electronic devices that generate repeating or non-repeating electronic signals and are used in designing, testing, troubleshooting, and repairing electronic devices.

AF Signal Generators

An *audio frequency signal generator* is a piece of test equipment that produces electronic waveforms at audio frequencies, roughly speaking from 20 to 20,000 Hertz or cycles per second. Depending on the unit they can produce various types of waveforms, the most common being sine and square waves. They find use in testing and troubleshooting of various types of electronic equipment including radio and audio amplifiers. Applications include frequency measurement (using headphones or an oscilloscope to compare the generated frequency with the frequency to be measured). A square wave output can be used to test amplifiers for frequency response and distortion.

RF Signal Generators

Radio Frequency or *RF signal generators* are capable of producing frequencies in the range used for radio. RF signal generators are used for servicing and aligning radio and television receivers. They often provide a facility to modulate the RF signal with an audio tone so that it can be heard on a radio receiver.

Function Generators

Function Generators produce signals with different waveforms, the most common being sine, square and triangle, but may support other types including arbitrary waveforms that the user defines. They typically support signals at audio frequencies or possibly up to low radio frequencies.
Sweep and *Alignment Generators* produce signals that sweep or change frequency at a specific rate and are useful for testing and adjustment of devices that use frequency modulation such as FM radio and television receivers, and for adjusting tuned circuits that need to have a particular response over a range of frequencies. The output is often displayed using an oscilloscope. A *Marker Generator* can produce signals that have visible marks at specific frequencies that can be observed on an oscilloscope.

Pulse Generators

Pulse Generators produce rectangular pulses at a specific frequency, pulse width, and possibly other characteristics such as rise and fall time. They are used for testing certain types of electronic devices and for scientific experiments.

Miscellaneous Signal Generators

There are other types of specialized signal generators, mostly notably those used for television servicing. Dot and bar generators produce known video signals that are used for adjusting television images. Typically they produce a number of different patterns. Geometric patterns such as dots and lines are used to adjust the image for linearity and distortion. Bar patterns are used to adjust gray scale and on color televisions, correct reproduction of color. Simple black and white bar generators, like the

BG-1, were offered by Heathkit starting in the early 1950s, switching to color bar generators like the CD-1 by the end of the decade. By the 1980s a color alignment generator offering 16 different patterns as well as color level controls was offered as a battery operated hand-held unit, the IG-5240.

Table 15 is a complete list of Heathkit signal generator models.

Kit IG-102
$31⁹⁵

Kit IG-72
$46⁹⁵

Wired IGW-72
$69⁹⁵

Factory Aligned Coil And Bandswitch Assembly . . . assures accurate tuning calibration.

Precision components . . . clean parts layout . . . well thought out circuitry . . . all combine to make a dependable instrument.

General Purpose RF Signal Generator Covers 100 kHz To 220 MHz With 2% Tuning Accuracy

• Covers 100 kHz to 220 MHz in six bands • Large accurately calibrated dial scales • Factory wired and aligned coil and bandswitch assembly • Modulated or unmodulated RF output • 400 Hz signal modulation and 400 Hz audio output for audio tests • High-level "RF Direct" output on Berkeley Physics Laboratory version

WIDE TUNING RANGE AND ACCURATE FREQUENCY CALIBRATION HAVE MADE THE IG-102 WIDELY IMITATED . . . and widely chosen too! The versatility and accuracy of this instrument, plus its low price, have put it into thousands of service shops and laboratories the world over. The value of a wide range test instrument and the additional manual entitled "How To Understand And Use Your Signal Generator" (see page 70) have caused the IG-102 and its sister the IGW-19 to be chosen by literally thousands of educational institutions, as well. Check the specifications. Chances are the IG-102 is the instrument you require.

400 Hz AM MODULATED OR UNMODULATED RF OUTPUT . . . OR 400 Hz AUDIO ALONE . . . are the signals available. The 100,000 microvolt RF output is controlled by both fixed-step and variable attenuators. The 400 Hz output is 10 volts under no-load conditions.

WIRED BERKELEY PHYSICS LABORATORY VERSION (IGW-19) . . . features the additional direct RF output jacks necessary for the Berkeley Physics Laboratory experiments, and a 3% tuning accuracy of the modified instrument. (Caution! This version of the instrument is not suitable for radio and TV servicing, since high-level RF radiates from the additional output jacks.) Order The IG-102 For The Best Value In General Purpose RF Generators . . . Bar None!

Kit IG-102, 7 lbs. **$31.95**
IGW-19, Assembled Educational Model, 6 lbs. **$65.00**

IG-102 SPECIFICATIONS—Frequency range: Band A, 100 kHz to 320 kHz; Band B, 310 kHz to 1.1 MHz; Band C, 1 MHz to 3.2 MHz; Band D, 3.1 MHz to 11 MHz; Band E, 10 MHz to 32 MHz; Band F, 32 MHz to 110 MHz. Calibrated harmonics: 100 MHz to 220 MHz. Accuracy: 2%. Output: Impedance, 50 ohms; Voltage, 100,000 uV. Modulation: Internal, 400 Hz, 30% depth; External, approx. 3 V across 50 k ohms for 30%. Audio output: Approx. 10 V open circuit. Tube complement: (1) 12AT7, (1) 6AN8. Power requirements: 105-125 or 210-250 VAC 50/60 Hz, 15 watts. Dimensions: 6½" W x 9½" H x 5" D.

Heathkit Switch-Selected Audio Generator . . . A Precision Sine-Wave Signal Source

• A near-perfect output signal waveform • Ideal for servicing or trouble-shooting high-fidelity equipment—a required instrument for many R & D applications • Switch-selected output frequencies—10 Hz to 100 kHz • Less than 0.1% distortion from 20 to 20,000 Hz • Panel meter monitors output with scales in volts and dB • Metered output level and switch-selected frequency are accurate to within ±5% from 10 Hz to 100 kHz • Manual includes detailed operating instructions and an analysis of the IG-72's circuitry

LESS THAN 0.1% DISTORTION THROUGHOUT THE AUDIBLE RANGE . . . better than ±5% accuracy of metered output level and switch-selected frequencies from 10 Hz to 100 kHz. Those are the hard and fast specifications of the IG-72. Here is an excellent instrument for R & D work where a low-distortion sinusoidal signal source is required. Switch-selected frequency control saves valuable time, lets engineers flick to the desired frequency without hairline adjustments. First and second digits and multiplier are switch selected from 10 Hz to 100 kHz (Tuning accuracy is specified from 10 Hz . . . distortion percentage from 20 Hz), and the output attenuator is switch selected in 10 dB steps with vernier adjustment on each range. An ideal instrument for production quality control.

HOW CAN A KIT BE A PRECISION INSTRUMENT? . . . many of our potential customers ask. Very easily! . . . if the circuit is designed through careful "know-how" and meticulous testing . . . if durable, precision components are chosen from well-known manufacturers . . . if the layout of components is designed for logical, straight-forward assembly . . . and if the assembly manual is written with insight and care. Very easily for you, the kit-builder, when you choose Heathkit instruments, since that is the heritage built into every Heathkit electronic instrument . . . the IG-72, for example!

Kit IG-72, 8 lbs., no money down **$46.95**
Assembled IGW-72, 9 lbs., no money down **$69.95**
Export model available for 115/230 VAC, 50-60 Hz; write for prices.
IG-72 SPECIFICATIONS—Frequency: 10 Hz to 100 kHz, switch selected, 2 significant figures and multiplier. Output: 6 ranges 0 to .003, .01, .03, .1, .3, 1 volts RMS into external 600 ohm load or with internal load into Hi-Z, 2 ranges 0 to 3, 10 volts RMS into a minimum of 10,000 ohms, —60 dB to +22 dB in 8 steps, —60 dBm to +2 dBm (0 dBm=1 mw into 600 ohms). Distortion: Less than .1%, 20 to 20,000 Hz. Tubes: (1) 6AU6, (1) 6CL6, (1) 6X4. Power: 105-125 or 210-250 VAC 50/60 Hz, 40 watts. Dimensions: 9½" W x 6½" H x 5" D.

Illustration 28: Two Signal Generators from the 1971 Catalog

Table 15: Heathkit Signal Generators

Model	Description	First Year	Last Year	Comments
AD-1309	Noise Generator	1985	1991	White and pink noise generator
AG-7	Signal Generator - AF	1951	1951	20 Hz to 200 kHz, sine/square wave
AG-8	Signal Generator - AF	1951	1957	18 Hz to 1 MHz, sine wave
AG-9	Signal Generator - AF	1956	1957	20 Hz to 110 kHz, sine wave
AG-9A	Signal Generator - AF			10 Hz to 110 kHz, sine wave
AG-10	Signal Generator - AF	1958	1962	20 Hz to 1 MHz, sine/square wave
AO-1	Signal Generator - AF	1951	1959	20 Hz to 20 kHz, sine/square wave
BG-1	Bar Generator	1953	1956	Horizontal and vertical bars for B&W TV
CD-1	Color Bar/Dot Generator	1957	1962	For television servicing
EF-3	Signal Generator Course			Includes test chassis. EF-3-2 includes IG-102
ES-600	Function Generator	1956	1956	For ES-400 computer
ETI-7020	Function Generator	1990	1990	With counter, assembled
EU-81	Function Generator	1971	1972	0.1 Hz to 1 MHz
EUW-27	Signal Generator - AF	1968	1970	20Hz to 1MHz, sine/square, assembled
FMO-1	FM Sweep Alignment Generator	1960	1967	90/100/107 MHz, 400Hz modulation, 10.7MHz sweep
G-1	Signal Generator - RF	1950	1950	150 kHz to 102 MHz
G-2	Signal Generator - Sine/Square			20 Hz to 20 kHz, sine/square wave
G-3	Sweep Generator	1948	1949	FM 10.7 MHz
G-4	Signal Generator - RF			150 kHz to 30 MHz
G-5	Signal Generator - RF	1950	1950	150 kHz to 102 MHz
IC-62	Color Generator			For television servicing
IF-1272	Signal Generator - AF	1977	1982	Low distortion
IG-14	Marker Generator	1968	1970	Sine/square
IG-18	Signal Generator - AF	1969	1975	Sine/square
IG-18A	Signal Generator - AF	1975	1977	Sine/square
IG-28	Color Generator	1969	1977	For television servicing, produces 12 patterns
IG-37	FM Stereo Generator	1968	1976	For alignment of FM radios
IG-42	Signal Generator - RF	1962	1979	Replaced LG-1
IG-52	TV Alignment Generator	1967	1972	For television servicing
IG-57	Post Marker/Sweep Generator	1968	1971	For television servicing
IG-57A	Post Marker/Sweep Generator	1970	1978	For television servicing
IG-62	Bar/Dot Generator	1967	1968	For television servicing
IG-72	Signal Generator - AF	1962	1977	10 Hz to 100 kHz
IG-82	Signal Generator - AF	1963	1968	Sine/square
IG-102	Signal Generator - RF	1963	1977	100 kHz to 110 MHz
IG-102S	Signal Generator - RF			Berkeley Physics Laboratory version of IG-102
IG-112	FM Stereo Generator	1964	1967	For alignment of FM radios
IG-1271	Function Generator	1975	1987	Sine/square/triangle, 0.1 Hz to 1 MHz

Model	Description	First Year	Last Year	Comments
IG-1272	Signal Generator - AF	1977	1983	5 Hz to 100 kHz, low noise
IG-1273	Function Generator	1977	1979	Sine/square/triangle, 0.3 Hz to 3 MHz
IG-1275	Function Generator	1977	1983	Sine/square/triangle, 0.3 Hz to 3 MHz
IG-1277	Pulse Generator	1984	1987	Variable pulse period, width, delay
IG-5218	Signal Generator - AF	1977	1990	Sine/square, to 100 kHz
IG-5228	Bar/Dot Generator	1977	1983	For television servicing, 12 patterns
IG-5237	FM Stereo Generator	1977	1979	For alignment of FM radios
IG-5240	Color Generator	1976	1984	Hand-held, battery operated, 16 patterns
IG-5242	Signal Generator - RF	1979	1979	Same as IG-42 but with "new look"
IG-5257	Post Marker/Sweep Generator	1977	1984	For television servicing
IG-5260	TV Alignment Generator	1989	1991	Hand-held, battery operated
IG-5280	Signal Generator - RF			310 kHz to 110 MHz
IG-5282	Signal Generator - AF	1977	1991	Sine/square, 10 Hz to 100 kHz
IGW-19	Signal Generator - RF			Berkeley Physics Laboratory, assembled version of IG-102S
IGW-47B	Signal Generator - Sine/Square			Berkeley Physics Laboratory, assembled, same as EUW-27
IGW-57A	Post Marker/Sweep Generator	1970	1978	Assembled version of IG-57A
IO-101	Vectorscope/Color Generator	1970	1977	3" vectorscope and color bar/pattern generator
LG-1	Signal Generator - RF	1953	1962	150 kHz to 30 MHz
LP-1	Linearity Pattern Generator	1955	1956	For television servicing
LP-2	Linearity Pattern Generator	1957	1957	For television servicing
RF-1	Signal Generator - RF	1960	1962	100 kHz to 110 MHz
SG-1	Signal Generator - RF			Square wave
SG-5	Signal Generator - RF	1950	1950	160 kHz to 50 MHz
SG-6	Signal Generator - RF	1951	1951	160 kHz to 50 MHz
SG-7	Signal Generator - RF	1951	1953	150 kHz to 150 MHz
SG-8	Signal Generator - RF	1956	1961	160 kHz to 110 MHz
SG-18A	Signal Generator - Sine/Square			Assembled version of IG-18
SG-57A	Post Marker/Sweep Generator	1970	1978	Assembled version of IG-57A
SG-1271	Function Generator			Assembled version of IG-1271
SG-1272	Signal Generator - AF			Assembled version of IG-1272
SG-1274	Function/Pulse Generator	1988	1992	2 MHz swept, assembled
SG-5218	Signal Generator - AF			Assembled version of IG-5218
SQ-1	Signal Generator - AF	1951	1957	Square wave
TS-1	TV Alignment Generator	1950	1950	For television servicing
TS-2	TV Alignment Generator	1950	1953	For television servicing
TS-3	Sweep Generator	1953	1954	For television servicing, 4 MHz to 220 MHz
TS-4	TV Alignment Generator	1956	1956	For television servicing
TS-4A	TV Alignment Generator	1957	1962	For television servicing

In-Depth: The IG-102 RF Signal Generator

In this section we'll look in detail at one of Heathkit's more popular pieces of test equipment, the IG-102 RF Signal Generator. It was offered from 1963 to 1977. The 1971 US Heathkit catalog lists it at a price of $31.95 while in 1976 it was $44.95.

The IG-102S was the Berkeley Physics Laboratory version which was identical except for an additional set of jacks for a high-level "direct" output needed for experiments in the laboratory. It should be noted that this version is not suitable for radio and TV servicing as high level RF radiates from the additional jacks. One could easily remove or disconnect the extra jacks in the laboratory version to turn it into an IG-102.

Both the IG-102 and IG-102S were kits. The IGW-19 was a factory assembled version of the IG-102S, selling for about $10 more than the IG-102 in 1971.

As part of their educational products line, Heathkit sold the EF-3 course "How to Understand & Use Your Signal Generator" which was intended for use with the IG-102 and included a test chassis. Model EF-3-2 was a bundle that included both the EF-3 course and the IG-102.

Marketed as "the best value in general purpose RF generators bar none", it was said to be the choice of thousands of educational institutions, service shops, and laboratories. The specifications, taken from the manual, are listed below:

SPECIFICATIONS

RF OUTPUT SIGNAL

Frequency Range	100 kc to 110 mc in six bands (bands A through F).
	100 mc to 220 mc (additional band of calibrated harmonics).
Frequency Accuracy	±2% (±3% for IG-102S/IGW-19 when RF Direct jacks are not used).
Output Voltage	.1 volt or higher
Output Impedance	50 ohms.
Internal Modulation	400 cps (30% modulation).
External Modulation	3 volt signal input for 30% modulation.

AF OUTPUT SIGNAL

Frequency	400 cps.
Output Voltage	10 volts (open circuit).

GENERAL

Front Panel Controls	Variable Frequency control.
	Band switch.
	Fine Attenuator control.
	Coarse Attenuator switch.
	External Modulation or AF Output control.
	Modulation switch.
Tube Complement	12AT7 RF oscillator.
	6AN8 amplifier and modulator.
Power Requirements	105 to 125 volts AC, 50/60 cps, 15 watts.
Cabinet Dimensions	6-1/2" wide x 9-1/2" high x 5" deep
Net Weight	4-1/2 lbs.

The unit is a pretty standard RF signal generator that can produce amplitude modulated or unmodulated RF signals in six overlapping bands suitable for AM, FM, TV, long wave, and shortwave receivers. The bands cover the following frequencies:

Band A: 100 – 320 kHz
Band B: 310 – 1100 kHz
Band C: 1 – 3 MHz
Band D: 3.1 – 11 MHz
Band E: 10 to 32 MHz
Band F: 32 – 110 MHz

Illustration 29 shows the front view of the unit. The commonly used IF frequencies of 455 kHz and 10.7 MHz are marked on the dial scale. An additional scale inside of band F is calibrated for the harmonics of band F from 100 to 220 MHz. Both the unit and the manual list frequencies in CPS (cycles per second), KC (kilocycles) and MC (megacycles), as was standard at the time.

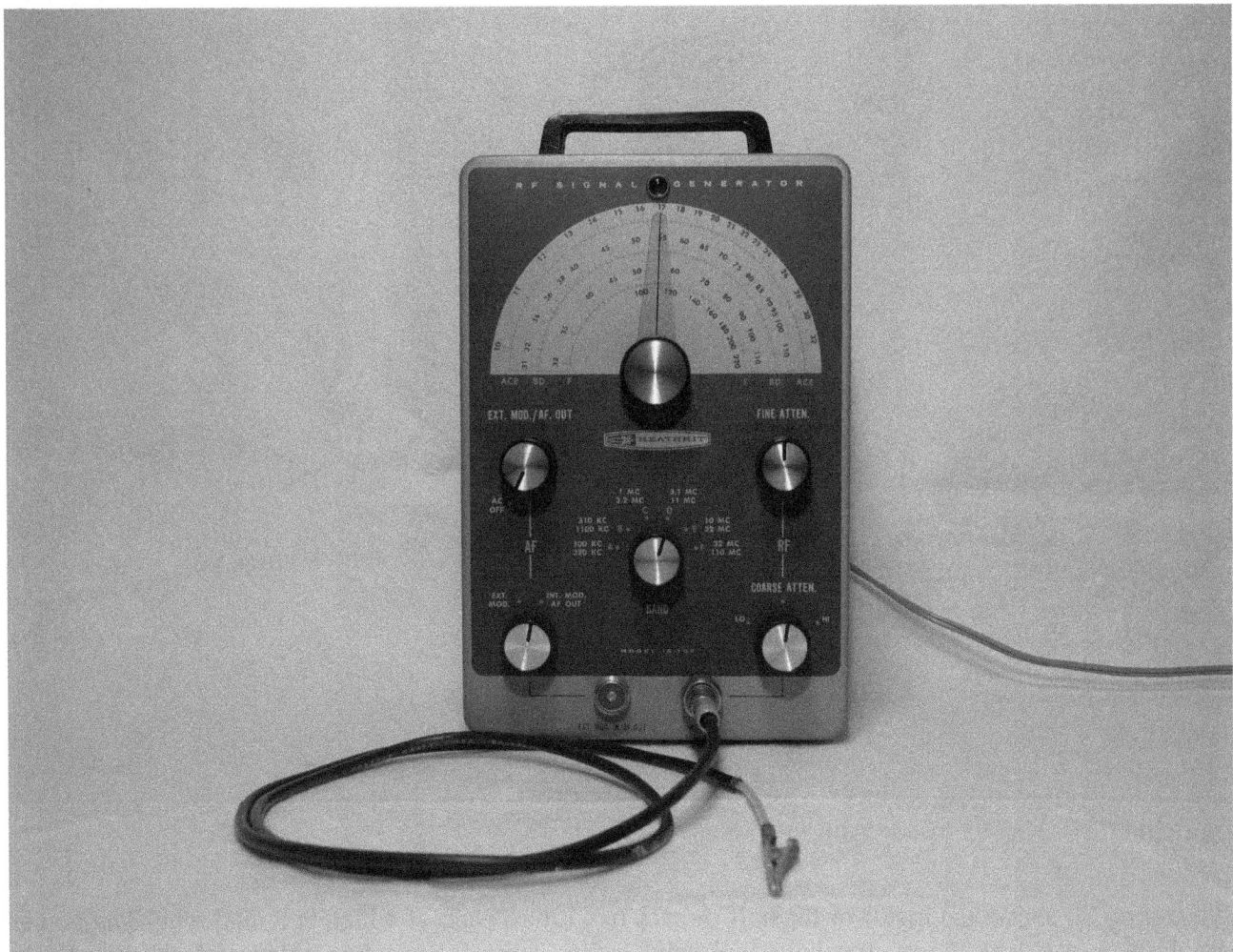

Illustration 29: IG-102 Front View

In the IG-102S/IGW-19 model the RF direct output banana jacks are located between the band switch and tuning dial.

The RF output level is controlled by fine and coarse attenuators but is not calibrated. The modulation frequency is fixed at around 400 Hz at about a 30% modulation level. The unit can accept an external modulation input. It can also directly output the 400 Hz audio signal.

Tuning uses a vernier drive on the dial. It used the same size and style of case as some other Heathkit instruments of the period like the IT-11 capacitor checker. It came with "microphone type" input/output connectors which are often replaced with more modern and widely available BNC connectors. The manual has about three pages covering applications for the generator such as AM radio alignment and signal injection for testing of TV and hi-fi amplifiers.

Illustration 30: IG-102 Inside View

Illustration 30 shows the inside of the unit. It uses two tubes, each of which is a dual tube. The power supply uses a silicon diode. The coils and band switch were factory assembled and aligned. Basic alignment could be done without instruments, using an AM radio and station of known frequency. If you had an accurate shortwave radio receiver that could be tuned to a frequency standard station like WWV you could adjust the coils for additional accuracy. Band F can be adjusted using an FM radio tuned to a station of known frequency.

I obtained my unit in 2013 from another local amateur radio operator. It was in good condition and did not need any adjustment to the alignment. It came with an original manual for the IG-102S laboratory version rather than the standard IG-102. Since my unit is an IG-102 it could not have been the original manual for this unit. It has the original microphone type connectors and a cable which appears to be original.

For the price, the IG-102 was a good value in a small, reliable generator with basic features needed for applications like radio receiver testing and alignment and Heathkit sold many of them over the years.

In-Depth: The TS-3 TV Alignment Generator

While the IG-102 was advertised as being suitable for FM radio and TV servicing, it was lacking some features needed to do full alignment of FM radios and television. Let's look at the Heathkit TS-3 Television Alignment Generator, an example of a specialized instrument specifically intended for television servicing.

The TS-3 is a sweep and marker generator offered in kit form from 1953 to 1954. It was the successor to the TS-1 and TS-2 models, and was followed by the TS-4. A 1953 Heathkit catalog lists it at a price of $44.50. The unit's specifications, as listed in the manual, are shown below:

SPECIFICATIONS

Frequency Range	4 mc to 220 mc continuous on four bands, all on fundamentals.
Output	Well in excess of .1 volt, regulated
Output Impedance	50 ohms terminated at end of output cable
Sweep	Operates upward only in frequency from base frequency. Sweep is across desired band rather than through each side from center.
Sweep Deviation	0-12 mc minimum up to 50 mc, depending on frequency. Sweep obtained electronically.
Fixed Frequency Marker	4.5 mc crystal, included with kit. Other frequency crystals may be quickly substituted if desired.
Variable Frequency Marker	19 mc to 60 mc on fundamentals, 57 mc to 180 mc on calibrated harmonics. Accuracy limited only by crystal accuracy.
External Marker	Any frequency can be mixed with crystal and variable marker oscillators to provide as many as three marker pips on one trace. Marker energy can be taken out from external marker connectors for separate application.
Attenuators	Step and "fine" controls for sweep oscillator, separate control for marker amplitude.
Blanking	Two-way blanking circuit incorporated to insure sharp oscillator cut-off.
Phasing	Wide range phasing control to allow easy trace centering.
Tube Complement	12AT7 – sweep oscillator and buffer 12 AT7 – variable and crystal marker oscillator 12AU7 – blanking clipper 6AQ5 – series high voltage regulator 6AU6 – regulator control 0A2 – control tube reference regulator 6X5 – rectifier
Cables	Output cable, 'scope horizontal cable and compensated 'scope vertical cable provided.
Power Requirements	110 volt AC 50/60 cycle, 60 watts
Dimensions	13" wide x 8 1/2" high x 7" deep
Net Weight	10 lbs.
Shipping Weight	18 lbs.

A sweep/marker generator is used for television alignment and testing. Television sets of that era had various tuned circuits that needed to be adjusted for bandwidth and proper frequency response curve shape. The basic technique is to inject a signal that changes (sweeps) in frequency and observe the behavior in the circuit using an oscilloscope. What is displayed is a graph of gain (vertical) versus frequency response (horizontal).

The term *sweep* refers to the radio frequency output that changes or sweeps over a range of frequencies. The generator can also produce *markers*, which are small vertical bars or "pips" that show up on the oscilloscope trace to mark a particular frequency. These allows one to check that the response at a specific frequency is correct. The TS-3 supports fixed, variable, and external markers: Fixed markers are at a frequency determined by a crystal plugged into a front panel socket. Variable markers appear at a frequency set by the **MARKER** dial (or one of its harmonics). An external marker can be produced by the input from an external signal generator. The feature known as *blanking* eliminates the return trace when the oscillator returns to the starting frequency. Without this, a double trace would be present on the oscilloscope, which would make the trace difficult to interpret.

Specialized sweep/marker generators like this were standard tools used for television servicing. This was still in the era of black and white television. As we will see in the next section, color TV added more requirements for test equipment. The TS-3 was one of a succession of TS models that had short lifetimes of just a few years each. This reflected that fact that television technology and techniques for TV servicing were changing rapidly at the time and television servicing equipment was continuously evolving.

This type of television servicing equipment is now obsolete for a number of reasons. Modern TVs typically no longer require alignment, the circuitry is fixed and/or aligns itself. Furthermore, analog television broadcasting (using the NTSC standard) has been phased out in most of North America and replaced by digital television, although most televisions still accept an analog input.

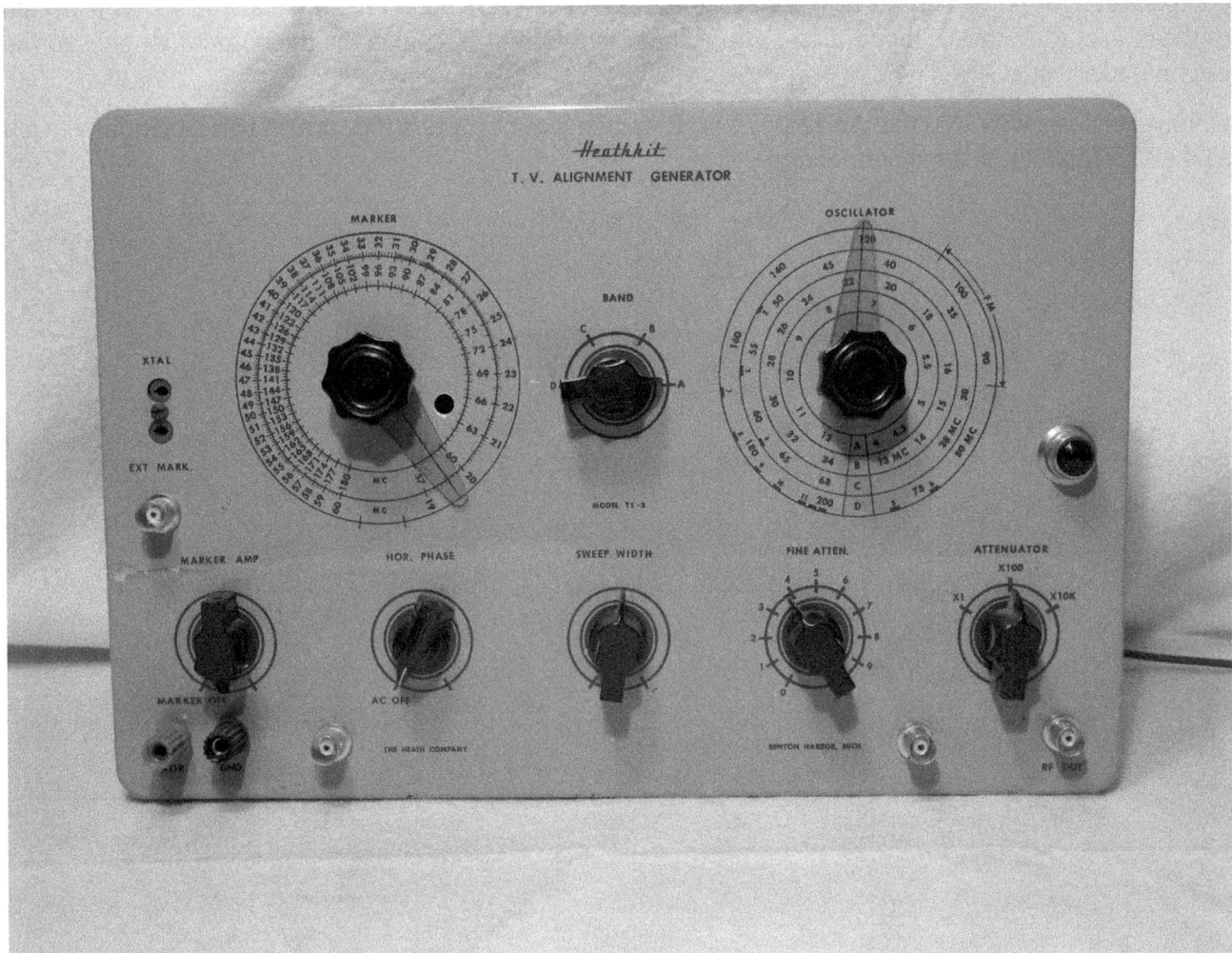

Illustration 31: TS-3 Front View

Illustration 31 shows the front panel of the TS-3. Let's briefly run through the controls and connectors. The BAND switch selects one of four frequency bands:

Band A: 4 – 12 MHz
Band B: 13 – 34 MHz
Band C: 28 – 6 MHz (including VHF television channels 2 through 4)
Band D: 75 – 200 MHz (including television channels 5 through 13 and and FM radio band)

Note that this unit predates the introduction of the UHF television channels 14 and up, and therefore does not support them. The **MARKER** control sets the position of a variable marker, from 19 to 60 MHz or 57 to 180 MHz using harmonics. The **XTAL** pins are a crystal socket for a fixed marker, with a 4.5 MHz crystal included. The unit can also accept an external marker signal using the **EXT. MARK.** connector.

The **MARKER AMP.** control sets the level of the variable marker or can turn it off. **SWEEP WIDTH** controls the amount of deviation of the sweep oscillator, from 0 to 12 MHz. **HOR. PHASE** compensates for phase shift in the receiver under test. The **HOR.** and **GND.** banana jacks typically go to the X axis (horizontal) input of an oscilloscope. The **RF OUT** connect is the output that goes to the circuit under test.

Output level is set by a coarse **ATTENUATOR** switch that has **X1, X100, and X10K** positions as well as the **FINE ATTEN.** attenuator control.

Illustration 32: TS-3 Inside View

Illustration 32 shows the inside of the unit. It was quite a complex kit to assemble with seven vacuum tubes and lots of point to point wiring. There are 20 pages of assembly instructions in the manual.

The sweep oscillator uses a device called an *Increductor Controllable Inductor* which varies its inductance with the current flowing through it. Some sweep generators of this era instead used mechanical systems which were complex, noisy, less reliable, and more expensive. The power supply circuit uses a selenium rectifier. This is an early type of solid-state rectifier that is not particularly reliable. They are sometimes known to fail spectacularly with a "distinctive" smell. Opinion varies on whether to replace them or leave them as is if they appear to be working okay. In general I would suggest to leave the rectifiers in place except for devices that you plan to use on a daily basis for long periods of time. In that case, consider replacing it with one or more modern rectifier diodes, but depending on the circuit, you may need to account for the different voltage drop of a modern silicon diode.

Alignment of the unit requires a VTVM or similar multimeter to adjust a calibration control to a known DC voltage. It also uses the 4.5 MHz crystal oscillator to adjust the sweep oscillator. This requires a signal tracer or 'scope with a demodulator probe. Television servicing applications for the TS-3 include sound and adjacent channel trap alignment, IF alignment, sound IF alignment, oscillator and RF alignment. It can also be used to align FM radio receivers.

The manual suggests as a full set of Heathkit instruments for television servicing, the TS-3, an oscilloscope such as the O-9 (with suitable probes), a VTVM such as the V-6, and optionally a grid dip meter like the GD-1B. A Bar Generator like the BG-1 is also recommended.

I purchased this particular unit on eBay in 2006. One of the risks of buying larger equipment like this on eBay is the possibility of damage in shipping. Unfortunately, this was one of the few cases I have had where I was unlucky and there was damage. A knob was broken, some jacks and some of the front panel were bent, and some parts were loose inside the unit. I was able to repair the damage, although the case still have some of the paint missing.

A number of modifications had been made to the unit by a previous owner. A filter choke had been added to the power supply (this is what came loose inside during shipping). The original 6X5 rectifier tube was replaced with a 5Y3GT and the circuit rewired accordingly. The power transformer was replaced (the new one has a 5 volt filament winding needed by the 5Y3GT).

If I had to guess I would theorize that at some time the original 6X5 rectifier tube shorted and overheated the transformer. The 6X5 tubes were known to do this, and there was evidence (black marks on the chassis) that could have been caused by the transformer burning. I think the owner then rewired it with a 5Y3GT tube (which are less prone to shorting) and a new transformer, and added a filter choke.

Other modifications included a hole drilled in the front panel to get at the inductor for the marker oscillator (presumably for calibrating it without opening the case). The RF OUT jack was changed from the original "microphone type" to a more modern BNC connector. Two additional jacks were also added. The circuit had been modified so that the marker output came out separately to a BNC jack rather than being mixed with the sweep output. A second jack duplicated the RF out, presumably so that a jumper cable between the jacks would add the marker output to the sweep output, when this was desired. Finally, it was missing the handle on the top of the case. These were sometimes removed (and subsequently lost) so that other equipment could be stacked on top.

The unit received a thorough going over including cleaning and replacing some parts. The end result is that it is in not too bad shape cosmetically and it works electrically. My intention was to use it for restoring some old 1940s or 1950s era televisions, and I hope to do so in the future. It is also useful for FM radio alignment and looking at tuned circuits in amateur radio equipment. As a straight signal generator it is less useful since the range of output frequencies are limited to 4 MHz and above.

In-Depth: The IG-57A TV Post-Marker/Sweep Generator

By the end of the 1960s, television in North America had evolved to add support for color and expanded from the original VHF channels 2 through 13 to add UHF channels 14 to 83. Television servicing equipment also evolved to support the new television technology. In this section we'll look at the IG-57A Post-Marker/Sweep Generator, a unit that was representative of this era. Here is a description of the unit taken from the introduction section of the assembly manual:

"The Heathkit Model IG-57A TV Post-Marker/Sweep Generator is a solid-state fifteen-crystal marker generator and electronic sweep generator for alignment of tuned circuits in color or black and white TV sets, and FM receivers.

The Post Marker Generator mixes one or more marker signals with the demodulated signal from the circuit being tested or aligned. The markers are sharp and well defined and will not alter or distort the response curve of the circuits involved. Therefore, the oscilloscope will show the actual waveshape of the circuit or device under test.

As many as six markers can be made to appear simultaneously on an IF trace. This enables you to adjust the IF circuits for proper waveshape and bandwidth in much less time than would be possible if you were to use the old variable marker system which must be reset and calibrated for each marker frequency

Markers are provided for color bandpass alignment; picture and sound carrier frequencies for channels 4 and 10; FM tuner, FM IF, and discriminator alignment; and television sound IF adjustments. Modulation at 400 Hz is provided for trap adjustment and for checking and adjusting FM tuners. Also provided are two variable voltage bias supplies, with switch-selector to provide positive or negative voltages.

The Sweep Generator has three linear sweep ranges. These ranges cover the sweep necessary for proper alignment of FM receivers and the TV tuned circuits in the sound IF, color bandpass, and video IF circuits, and for proper overall RF/IF response.

The diode modulator combines the frequencies of the post-marker and sweep generators to permit amplitude modulation of the picture carrier frequency by a low frequency sweep signal. This method of sweep alignment, called Video Sweep Modulation, permits observation of the overall color bandpass response, including the effect of the video detector load circuitry.

Other features include a blanking switch, trace reversing switch, and a phase control so the markers will appear as shown in the the waveforms in the set manufacturer's alignment instructions regardless of the oscilloscope you use.

All of these features combine to provide you with a versatile, accurate, and attractive test instrument that is designed for long and dependable service at minimum cost."

The IG-57A was offered from 1970 to 1978. The cost was $135.00 in 1971 and $164.95 in 1976. It was also available as a factory assembled version designated at various times as the IGW-57, IG-5327, and SG-57A. The assembled version had slightly different color styling and typically sold for about 50% more than the kit. The predecessor to the IG-57A was the almost identical IG-57 model.

The specifications of the unit are listed below:

SPECIFICATIONS

Marker Frequencies . . . . 100 kHz
     Crystal Controlled, ±.01% . . . . 3.08 MHz, 3.58 MHz
                                  4.08 MHz, 4.50 MHz
     Crystal Controlled, ±.005% . . . . . . 10.7 MHz, 39.75 MHz, 41.25 MHz,
                                  42.17 MHz, 42.50 MHz, 42.75 MHz,
                                  45.00 MHz, 45.75 MHz, 47.25 MHz, 67.25 MHz,

	193.25 MHz
Modulation Frequency	400 Hz
Input Impedance	External MKR/SWP – 75 Ω,
	Trace Input – 220 kΩ,
	Attenuator – 75 Ω.
Output Impedance	Marker Out – 75 Ω,
	Scope Vert – 1 kΩ,
	Sweep Output – 75 Ω,
	Attenuator – 75 Ω,
	IF/RF VSM – 75 Ω.
Bias Voltage	Positive or negative 15 volts DC at 10 milliamperes.
Type of Marker	Birdie.
Controls	Two individually adjustable Bias controls,
	Marker/Trace – dual concentric,
	Sweep Width/Sweep Center – dual concentric,
	Marker Out – concentric with Sweep Range switch,
	Phase.
Switches	Rocker Type – separate switch for each of the above listed frequencies,
	Blanking, On/Off,
	Trace Reverse,
	Modulation On/Off,
	Power On/Off,
	Bias + or -.
Transistor-Diode Complement	18 – 2N3692 transistors,
	6 – 2N2293 transistors,
	1 – 2N3416 transistor,
	2 – 40245 transistors,
	3 – silicon diode rectifiers,
	3 – crystal diodes,
	1 – 13.6 volt zener diode,
	1 – 20 volt zener diode.
Sweep Frequency Ranges and Output Voltage	LO Band – 2.5 to 5.5 MHz±1 dB at 0.5 volt RMS minimum, fundamentals, with 10.7 MHz on harmonics.
	IF Band – 38 to 49 MHz ±1 dB at 0.5 volt RMS minimum, fundamentals,
	RF Band – 64 to 72 MHz ±1 dB at 0.5 volt RS minimum, fundamentals, with 192 to 198 MHz on harmonics.
Attenuator	Total of 70 dB of attenuation in seven steps – 1 dB, 3 dB, 6 dB, 10 dB, 10 dB, 20 dB, 20 dB.
Power Requirements	120 volts, 60 Hz AC at 4.5 watts.
Dimensions	13-3/8" wide x 5-1/2" high x 12" deep.
Net Weight	14 lbs.

Illustration 33 shows the front panel of the unit. I won't describe the 30 or so controls and jacks on the front panel; if you are interested in the details you can download a copy of the assembly manual which runs more than 100 pages in length.

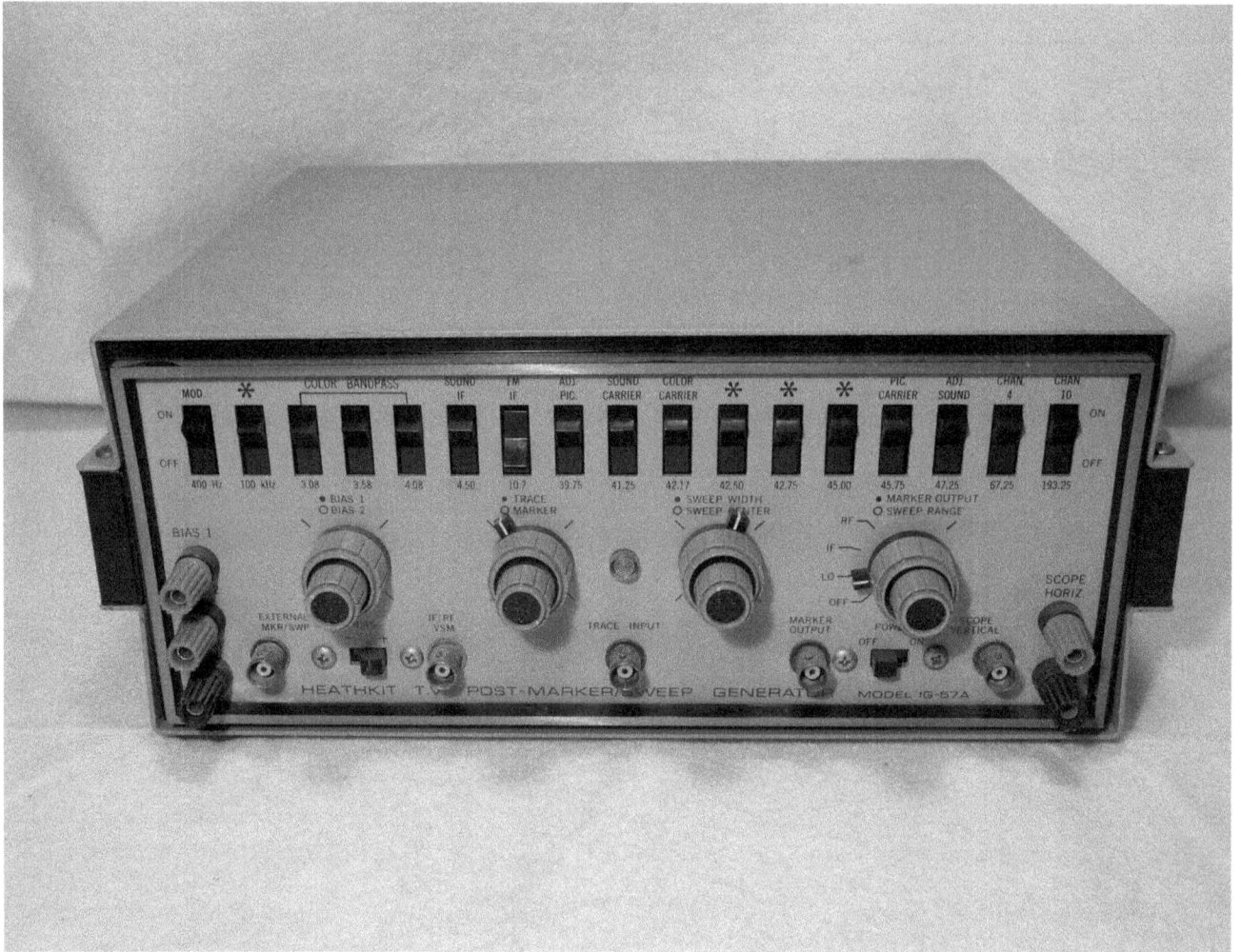

Illustration 33: IG-57A Front View

The term "post-marker" in the product name refers to the use of post-injection marker circuitry, where the markers are added to the sweep signal after the signal passes through the unit under test, making the marker "pips" cleaner and more visible. The alternative is pre-marker injection which some sweep generators used, and is considered an inferior design.

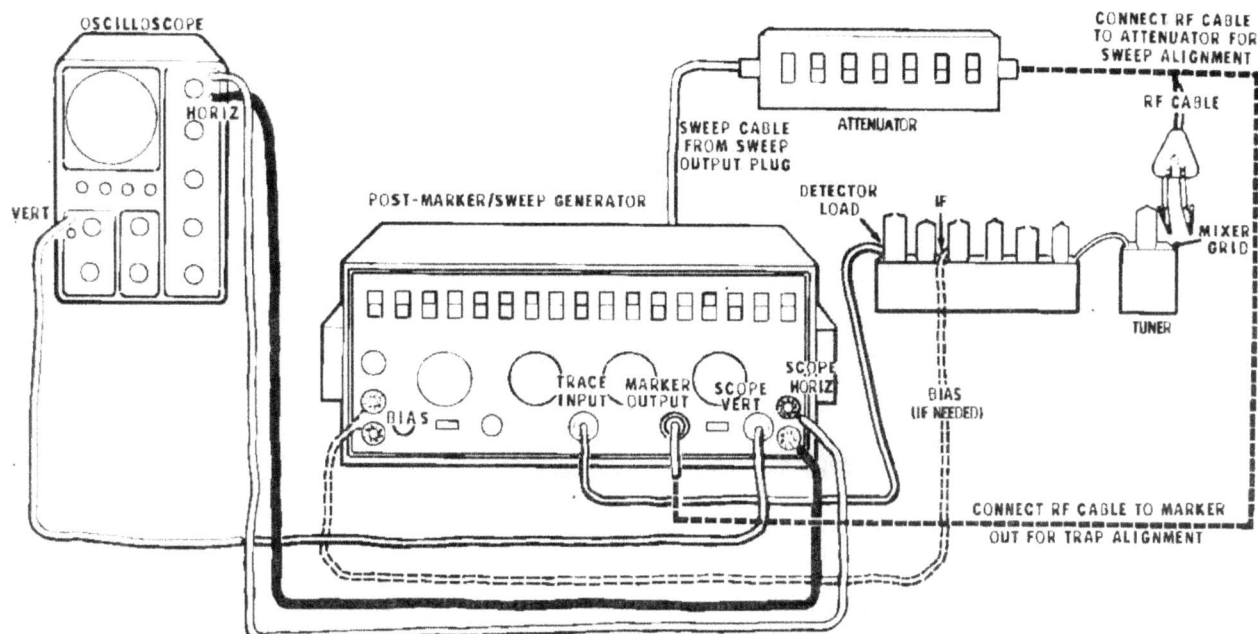

Illustration 34: IG-57A Connections for IF Sweep and Trap Alignment

Illustration 34 is a diagram from the manual showing a typical setup for television alignment. The unit consisted of the generator itself as well as a separate attenuator box and four coaxial cables with test leads and demodulation probe. The manual has over twenty pages of applications including procedures for aligning specific models of Heathkit televisions and generic procedures for television and FM radio alignment.

The unit was quite a complex kit to build with approximately 65 pages of assembly instructions in the manual. Illustration 35 shows a view of the inside. Construction is on three printed circuit boards with additional point-to-point wiring and a significant number of mechanical parts. The heart of the sweep generator is a controllable inductor, an electrical device in which the inductance of several oscillator coils is determined by the current in the control winding. Test and calibration of the unit requires a VTVM and oscilloscope but no other instruments as it uses its own crystal oscillators for frequency calibration.

As compared to the TS-3 described earlier, the IG-57A offers more features needed for servicing then-current television receivers with support for color and UHF channels. It also supports features for FM radio. It is all solid-state, and as such has much more complex circuitry, more densely packed on printed circuit boards. It has a more modern styling and color of the case and controls and uses the modern BNC type connectors.

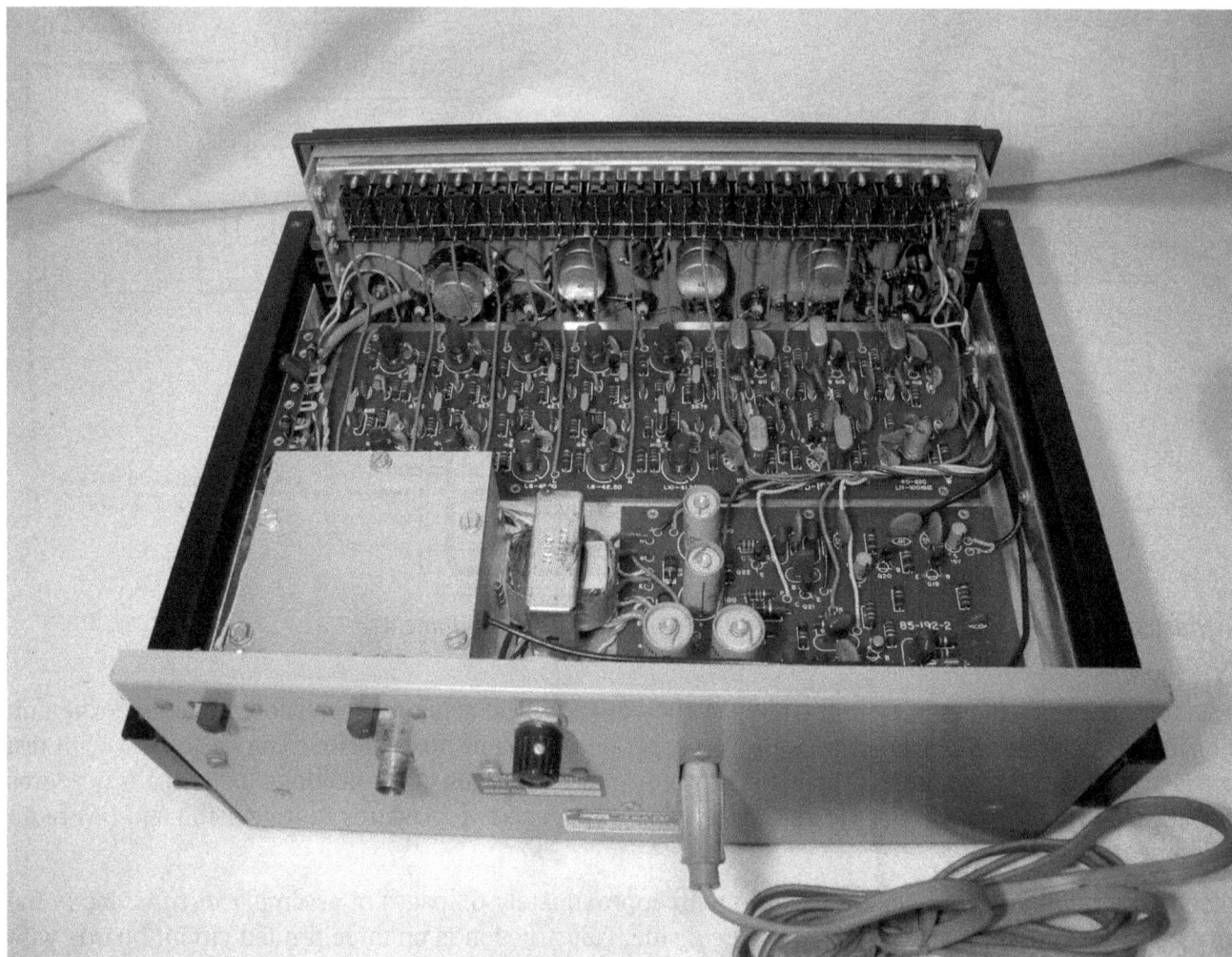

Illustration 35: IG-57A Inside View

I obtained my unit on eBay. It was being offered for a low price, but was missing the cables and attenuator box. I was able to find a full manual on the Internet. I also have an attenuator box with similar features that can be used with the unit. No real restoration was needed. I gave it a thorough cleaning and checkout and it seemed to be working fine.

In summary, the IG-57A was an example of a piece of test equipment that was state of the art in the early 1970s. It provided the features needed for television and FM radio alignment for receivers of the time and was offered for a price that was economical when compared to other commercial products. Heathkit claimed that the unit was used by thousands of technicians and servicemen. In conjunction with a VTVM and oscilloscope, it allowed one to align virtually any B&W or color receiver on the market at the time.

Chapter 10: Tube Testers and Checkers

At one time, tube testers could be found in almost every drug store in North America. The average person was capable of taking tubes out of a radio or TV, testing them at the drug store, and replacing them if they tested bad. When electronics went all solid state, this became a thing of the past. Because tubes had a limited lifetime and were relatively expensive, they were designed to be easily replaced. A tube tester could determine if a tube was functioning or not. The consumer tube testers were large and provided storage for a selection of new tubes. Tube testers were also sold that were aimed at engineers, manufacturers, radio and television repair technicians, and hobbyists. Heathkit made some models of this type. This book *Tube Testers and Classic Electronic Test Gear* goes into great detail on how tube testers worked and the different models that were on the market. Essentially there were two major general types of testers.

Vacuum tubes rely on an effect called "thermionic emission" where electrons can flow in a vacuum when emitted from a heated element called a cathode. The heating is performed by an element called a heater which is similar to a light bulb filament. The most basic testers can do an emission test that verifies that current can flow through the tube. On all but the very lowest cost testers a meter usually gives some indication of the level of emission. Most testers will also test for shorts between the elements. Typically this is done first before an emission test. This type of tester is usually referred to as a *tube checker*. Some tube checkers were even simpler and only tested that the tube's heater filament was not open.

Most tubes can amplify a signal. A measure of the tube's ability to amplify is known as its transconductance. A more sophisticated tester can measure the tube's actual transconductance value. Because of the complexity of doing this, most testers perform a simplified version of this test, often calling it some similar term such as *mutual conductance* or *dynamic conductance.*

The switch settings for testing a tube typically came from a *roll chart* which was integrated into the tester. Over time the roll charts were regularly updated to correct errors and add new tube types as they came on the market. Companies that sold tube testers, including Heathkit, typically offered yearly subscriptions for updates to the roll charts for a fee.

In old tube-based equipment a bad tube is actually one of the *least* likely causes of failure. Unless a tube is physically broken, much more likely problems are electrolytic capacitors, leaky wax paper capacitors, or carbon resistors that have changed value. After all these years, tubes are still available, both used and what is referred to as NOS (New Old Stock - old tubes that have never been used). Some tubes are still being manufactured in Russia and China for markets like electric guitar amplifiers and amateur radio transmitters. There are only a few tubes that are now hard to find.

Getting back to the drugstore tube tester, it is sometimes said that the purpose of tube testers was to sell tubes. These testers were often calibrated so that many tubes would test as questionable or weak to get the customer to purchase a new tube. Because most tube testers don't measure true transconductance, they only give an approximate result. The best test for a tube is the circuit it runs in. Don't throw out a tube because it reads weak on a tester, it may work fine in a radio. Discard it if it is physically broken, has a burnt out heater, or shorts (and even shorts are okay in some cases if the elements are not used).

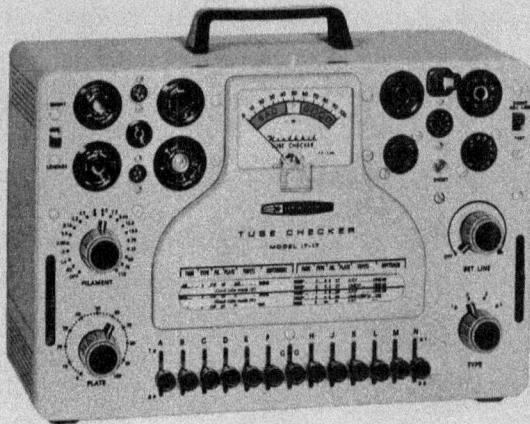

TT-1A
$179⁹⁵

IT-17
$54⁹⁵

Low Cost Heathkit Tube Checker... Color Styled To Match New Heathkit Instruments

• Fills most daily service requirements • Tests most tube types including new compactrons, nuvistors, novars, and 10-pin miniatures • Multi-colored "Bad-?-Good" meter scale • Tube-data chart is built-in—can't be misplaced • Features constant-tension free-rolling roll chart mechanism • Compact styling with carrying handle • Color-coded wiring harness simplifies point-to-point wiring • Requires AC VTVM for accurate alignment

AN ESSENTIAL INSTRUMENT FOR RADIO & TV SERVICING. Saves valuable time by eliminating doubt and guesswork as to tube quality before further circuit investigations are made. Tubes are quickly tested for quality on the basis of total emission, for shorts, leakage, open elements, and filament continuity. A fast, positive evaluation of tube quality. The relative quality is read on a multi-colored "bad-?-good" meter scale and a neon bulb indicator shows filament continuity and leakage or shorts between elements.

FLEXIBLE OVERALL OPERATION. A wide range of filament voltages is available through the front-panel control switch. Cathode current is adjustable. Meter sensitivity is variable for adjustment to individual tube characteristics. And individual switches matrix the tube base contacts. Thirteen 3-position lever switches allow each tube element to be checked independently for open or shorted conditions. These are features that protect your IT-17 against obsolescence . . . simplify the addition of new tube types to the chart tables. (See note below for replacement roll charts.)

SPECIAL ROLL CHART MECHANISM. The free-rolling chart mechanism featured in the IT-17 is spring-loaded to keep the chart taut as it is examined throughout its entire length. Chart and meter are both illuminated for easy reading of all tube test data.

EASY TO BUILD. Neat, professional wiring and simplified assembly is assured with the color-coded wiring harness and easy-to-follow instructions supplied. Order your Heathkit IT-17 now for dependable test facilities at low cost.

Kit IT-17, 12 lbs...no money down......................$54.95

IT-17 SPECIFICATIONS—Tests: Emission, short, leakage, open element, filament continuity. Sockets: 4-pin, 5-pin, 6-pin, 7-pin combination and pilot light, 7-pin miniature, octal, loctal, 10-pin miniature, 9-pin novar, 9-pin miniature, 12-pin compactron, 5-pin nuvistor, 7-pin nuvistor. Meter: 1mA BAD-?-GOOD scale, illuminated. Line voltage adjustment: Step type. Roll chart mechanism: Constant tension, free-rolling, thumbwheel operation, illuminated. Filament voltage: .63, 1.4, 2, 2.35, 2.5, 3.15, 4.2, 4.7, 5, 6.3, 7.5, 9.45, 12.6, 19.6, 25, 32, 50, 70, 110 V AC. Element test voltages: 30, 100, 250 V AC. Power requirements: 105-125 V AC, 50/60 Hz. Dimensions: 13" W x 8½" H x 5½" D.

NEW SUBSCRIPTION SERVICE FOR REPLACEMENT TUBE DATA ROLL CHARTS
Updated replacement roll charts for all Heathkit tube checkers can be purchased either singly or on a subscription basis which assures a new roll chart plus two supplements per year. Prices are $3.50 per chart; $5.00 for the new chart plus two supplements during the following year. Write to: Heath Tube Data Service, P.O. Box 377, Hewlett, Long Island, New York 11557

Heathkit Mutual-Conductance Tube Tester Tests Gm For Most Critical Tube Evaluation

• An outstanding array of features • Indicates Gm to 24,000 micromhos • Ultra-sensitive grid current test • Direct reading ohmmeter leakage test • Constant-current heater supplies • Built-in calibration circuit for high accuracy • Built-in roll-type tube data chart • Built-in adapter for testing compactron, nuvistor, novar, and 10-pin miniature tube types • Handsome leatherette covered carrying case.

NO FINER TUBE TESTER IN KIT FORM ANYWHERE. Versatile switching facilities accommodate virtually all tube types. Sure protection against obsolescence. The TT-1A provides quick and accurate tests of mutual conductance (Gm) from 0 to 3,000 micromhos, with multipliers to extend this range to 24,000 micromhos. In addition there is an ultra-sensitive grid current test, a life test, hybrid tube test, a direct-reading ohmmeter leakage test, and more. A disconnect switch removes all voltages from the selector switches to protect both tubes and tester against possible damage during setup. "Cathode buss," "plate buss," and "grid buss," switches allow testing of both sections of a twin triode individually, with only one group of selector switch settings for extra operating convenience.

SPECIAL FEATURES TO ASSURE YOU YEARS OF SERVICE. Constant-current heater supplies protect against obsolescence by supplying the correct current for tubes requiring a wide range of heater voltages. A switch-operated calibrating circuit designed into the TT-1A assures continued high accuracy for the life of the tester . . . no standardized tubes required . . . no extra voltmeters or other instruments. Chosen by professionals in virtually every phase of the electronics industry, the TT-1A offers sophisticated, thorough testing facilities at low cost.

Kit TT-1A, 33 lbs...no money dn. $179.95

TT-1A SPECIFICATIONS—Power requirements: Voltage: 105-125, 60 Hz AC. Watts: 10-60 (dependent upon tube under test). Plate supply (silicon rectifiers): DC volts: 26, 90, 135, 225, variable 80 to 200. (Separate DC supply for space charge grids.) AC volts: 2Ω, 45, 177. Bias supply (silicon rectifier): Low range: 0 to negative 5 volts DC. High range: 0 to negative 20 volts DC. Signal voltages: 2, 1, .5, .25 volts AC, 5000 Hz. Filament supply: Voltages: .65, 1.1, 1.5, 2, 2.5, 3.3, 5, 6.3, 7.5, 10, 13, 20, 27.5, 35, 47, 70, 115. Currents: 300, 450, 600 mA. (Note: Filament voltage is reduced 10% during life test.) Testing circuits: Gm: mutual conductance for amplifiers, 0-24,000 micromhos. Emission: rectifiers and diodes. Leakage: direct reading ohmmeter. Grid current: ¼ microampere sensitivity. Voltage regulator: firing voltage and regulation tolerance. Low power thyratron: grid characteristics, conduction capabilities. Eye tubes: control grid characteristics. Meter: AC: 1000 ohms/volt (1 volt full scale), DC: .89 ma full scale. Scales: Gm, 0-3000 micromhos. VR test volts, 0-200. Leakage, 0-10 megohms. Diodes, OK. Rectifiers, OK. Line check arrow at midscale. Tube complement: (1) 3A4 oscillator and (1) 12AV6 meter control. Calibration circuit: Built-in, switch operated. Socket accommodations: 4-pin, 5-pin, 6-pin, 7-pin combination and pilot light, 7-pin miniature, 7-pin subminiature, 8-pin subminiature, octal, loctal, 9-pin miniature, 10-pin miniature, 5-pin and 7-pin nuvistors, 9-pin novar, 12-pin compactron and blank. Line voltage adjustment: Continuously variable. Roll chart mechanism: Constant tension, free rolling, dual thumbwheel operated, illuminated. Dimensions: Cabinet (outside): 17¾" W x 13½" H x 8" D. Panel & chassis: 17" W x 12¾" H x 5½" D.

Illustration 36: IT-17 and TT-1A Tube Testers from 1971 Catalog

A tube tester is a impressive instrument with lots of knobs and switches but is not strictly necessary unless you have a need to test many tubes. When I restore an old radio or other equipment, if it works well after replacing the capacitors and out of value resistors, I usually don't bother to test the tubes. I also don't stock up on old tubes. Hundreds of different tubes were manufactured. I rarely need a new tube, and if I do it is invariably one I don't have, so I only buy tubes when needed. That being said, over the years one tends to acquire tubes, and I have a number in my junk box.

However, if you have an old piece of equipment that is important, you might want to purchase a spare set of tubes for it for future replacement. The stock of old tubes won't last forever.

Heathkit sold a number of tube checkers and testers, starting with the TC-1 in 1949 and followed later by TC-2 and TC-3 models which were very similar. Later units followed Heathkit's more standardized product naming scheme, the IT-21, IT-17, and IT-3117.

The company Weston made respected tube testers and for a period of time both Heathkit and Weston were divisions of the Daystrom company. The Weston 981-3 mutual-transconductance tester was adapted to become the Heathkit TT-1, a more sophisticated tester than the earlier models.

To test picture tubes (CRTs), Heathkit provided an adaptor for some testers. They also made some dedicated tube testers for CRTs, often called "rejuvenators" as they could sometimes fix shorts or improve the performance of old tubes. Table 16 lists the models of tube testers and checkers offered over the years.

Table 16: Heathkit Tube Testers and Checkers

Model	Description	First Year	Last Year	Comments
225	Adaptor			TV Picture Tube Test Adaptor for tube checker
355	Adaptor			TV Picture Tube Test Adaptor for use with TC-2
CC-1	Tube Tester - CRT	1956	1960	Emission type
IT-17	Tube Tester	1967	1976	Emission type, shorts, leakage,opens, filament test
IT-21	Tube Tester	1961	1967	Replaced TC-3
IT-3117	Tube Tester	1977	1981	Same as IT-17
IT-5230	Tube Tester	1976	1990	CRT tester and rejuvenators
ITA-5230-1	Adaptor	1976	1990	CRT socket adaptor for IT-5230
TC-1	Tube Tester	1950	1953	Emission tester
TC-2	Tube Checker	1953	1959	Emission tester, wood cabinet
TC-2P	Tube Checker	1953	1958	Portable version of the TC-2
TC-3	Tube Checker	1959	1962	Emission tester
TT-1	Tube Tester	1960	1961	Mutual Conductance tester
TT-1A	Tube Tester	1962	1967	Mutual Conductance tester

In-Depth: The IT-17 Tube Checker

In this section we'll take a detailed look at the Heathkit IT-17 Tube Checker. This model was offered from 1967 to 1976. It was relabeled as the IT-3117 in 1977 and sold until 1981. A 1971 catalog lists it at $54.95 and $99.95 in 1976. It was only offered as a kit.

This is an emission type tester, billed as a tube *checker* rather than a tube *tester*, where the tubes are tested for merit by connecting all the grids to the plate, operating the tube as a rectifier, and comparing the current flow to a predetermined value for the tube type, indicating the tube as good or bad on a meter. Typically Heathkit also offered a more expensive mutual conductance type tube tester at a higher price. In 1971 a TT-1A Mutual-Conductance Tube Tester on the same catalog page as the IT-17 was selling for $179.95.

Like most tube testers, settings for testing specific tubes are listed on a "roll chart" contained within the unit itself and visible through a window. To accommodate new tube types, Heathkit offered a tube chart subscription service, either single charts or a subscription of two supplements per year. In 1971 the cost was $3.50 per chart or $5.00 per year. Apparently this was handled by a third party as the catalogs list an address in Long Island, New York to contact rather than the Heath company. The roll charts were compatible between some testers from other manufacturers so you could potentially use the same roll charts (in some cases changing numbered levers to corresponding letters).

The specifications for the IT-17, taken from the manual, are below:

SPECIFICATIONS

Tube Socket Accommodations . . . . 4 - pin.

    5 - pin.

    5 - pin Nuvistor.

    6 - pin.

    7 - pin combination and pilot lamp.

    7 - pin miniature.

    7 - pin Nuvistor.

    8 - pin octal.

    8 - pin loctal.

    9 - pin miniature.

    10- pin miniature.

    12- pin Compactron.

Controls . . . . FILAMENT VOLTAGE

    SET LINE

    TYPE

    PLATE

Element Test Voltages . . . . .30, 100, and 250 volts AC.

Filament Voltages . . . . .63, 1.4, 2, 2.35, 2.5, 3.15, 4.2, 4.7, 5, 6.3, 7.5, 9.45, 12.6, 19.6, 25, 32, 50, 70, and 110 volts AC.

Roll Chart Mechanism . . . . Constant tension, free rolling, thumbwheel operated, illuminated.

Line Voltage Adjustment . . . . .Step type.

Meter . . . . 1 milliampere full scale, BAD - ? - GOOD, and Filament continuity

Tests Available . . . . Emission, Short, Leakage, Open Element, and Filament continuity.

Power Requirements . . . . 105-125 volts 50/60 cps AC.

Dimensions . . . . 13" wide x 8-1/2" high x 5-1/2" deep.

Net Weight . . . . 9 lbs.

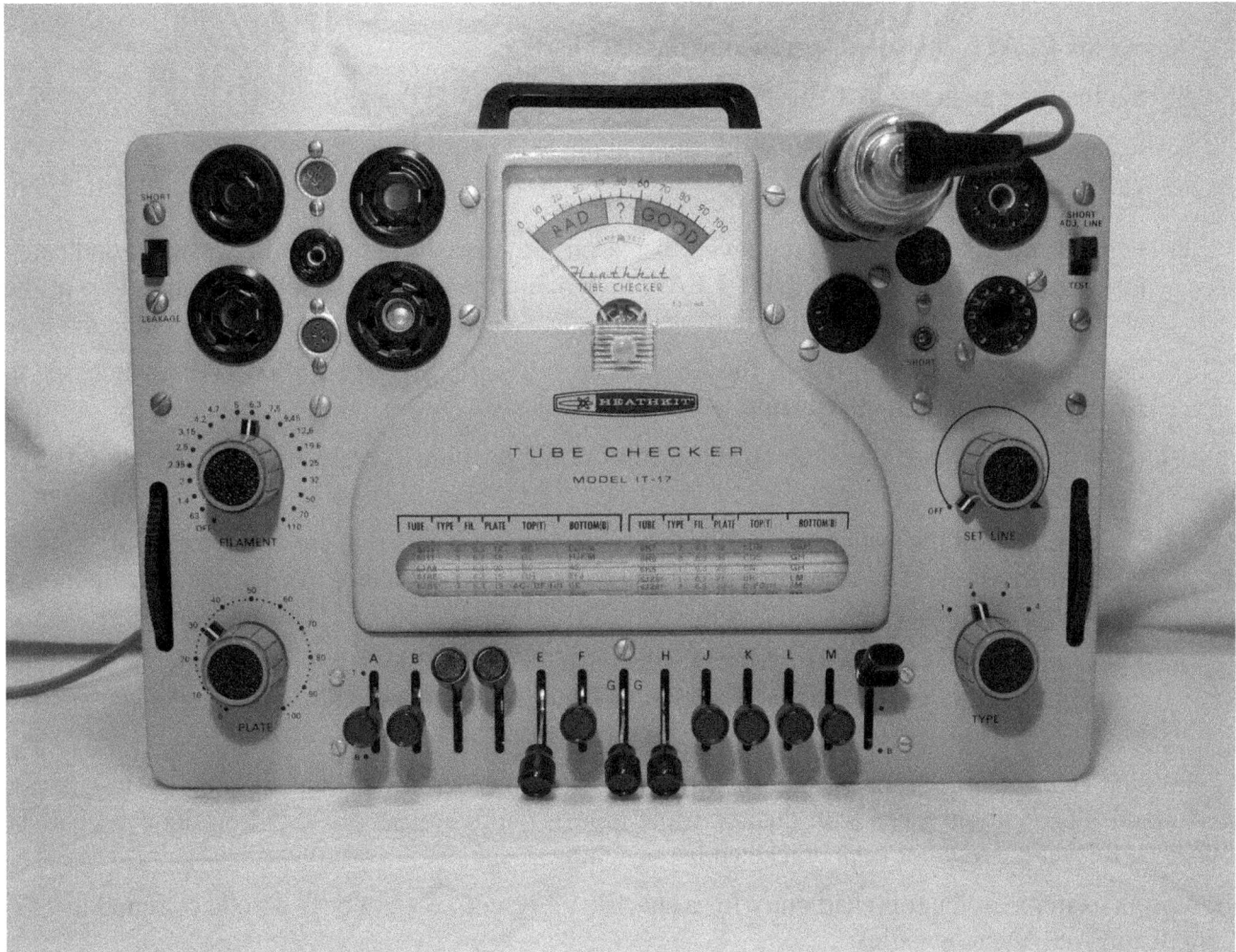

Illustration 37: IT-17 Front View

The manual covers the different types of tube testers and why emission testing (like this one) is a good tradeoff between test accuracy and the cost and complexity of the tester. Illustration 37 shows the front panel of the tester with a tube under test.

With sockets for 12 different tube types, it was better than some competing testers which didn't support some of the older types. You could also test pilot lamps using the center of the seven pin socket and setting the filament voltage to the appropriate voltage for the lamp. This is handy for testing bulbs like the 6.3 volt #47 type used in many old radios.

The meter is calibrated in arbitrary units from 0 to 100 with sections indicated as BAD, ?, or GOOD. The procedure for testing tubes is as follows:

1. Find the tube type on the roll chart. Turn the **SET LINE** control until the meter reading is within the **LINE TEST** block (this compensates for varying AC line voltage).

2. Set the **TYPE** switch as per the roll chart.

3. Set the **FILAMENT** selector to the voltage listed on the roll chart.

4. Set the **PLATE** control as specified in the roll chart.

5. Set the lever switches to **T** (top) or **B** (bottom) as per the roll chart.

6. Connect the test clip to the tube grid cap, if applicable.

7. Insert the tube in the appropriate socket and readjust the **SET LINE** switch if needed.

8. Perform Shorts test - move the levers listed on the chart in light type between the indicated positions and center. The **SHORT-LEAKAGE** switch is set to **SHORT**. A steady glow of the neon lamp for any switch setting indicates a short.

9. Perform Leakage test - repeat the above step but with **SHORT-LEAKAGE** switch set to **LEAKAGE**. An illuminated lamp will indicate leakage.

10. Quality test – Move the slide switch to the **TEST** position. The meter will indicate GOOD, BAD, or ?.

11. Open test - with the switch in the **TEST** position, move levers that are at the top to the bottom position and back. You should observe a decrease in meter reading. If not, the corresponding tube element is likely open.

12. Filament check – set the **FILAMENT** selector to .63 volts. Move the levers listed in dark type through the other two positions. The **SHORT** indicator lamp should light if the filaments are okay.

If the tube fails the shorts or leakage test it is usually considered bad and the quality test is not performed. For multiple tubes (e.g. dual or triple tubes), multiple tests are listed on the roll chart in brackets. Repeat the tests for each bracketed list of settings.

Here is an example of the roll chart entry for a specific tube, a 50L6 octal base tetrode common in "All American Five" AM tube radios:

TUBE TYPE	FIL.	PLATE.	TOP(T)	BOTTOM(B)
50L6	3	50	21	CDE	GH

If a tube is not listed in the chart, you can follow the generic instructions in the manual given the tube's base diagram. You need a known good tube of the type (or, preferably, at least three) to determine what **PLATE** control setting is needed for the tube to show in the middle of the **GOOD** range.

Illustration 38: IT-17 Inside View

Illustration 38 shows the inside of the unit. The circuit has only a few components, mostly switches and wiring. A power transformer has taps for all the filament voltages and high voltages. Other components include the neon lamp, 1 mA meter, and a handful of resistors and capacitors. The wiring harness came pre-assembled and color coded to make assembly easier. A VTVM is recommended for testing the assembled unit.

When looking at a used unit, check if the power transformer is bad. If so, it is probably not feasible to fix as it is unique to the tester. The meter is also hard to replace other than with a unit that likely does not match the same size and dial scale. The IT-17 uses a copper oxide rectifier which can fail over time. Replacements for this part are hard to find. The IT-3117 used a 1N191 germanium diode which you may be able to find and use instead. Alternatively it can be replaced with a modern diode, preferably of germanium type.

Manuals and roll charts can be found on the Internet. If your roll chart is missing or destroyed you could create a replacement with a little effort and ingenuity (e.g. cutting and gluing printed pages together).

I purchased my IT-17 from a local seller on Kijiji. The unit was working fine when received. I gave it some cleaning and a full checkout. The only modification was that at some point someone had repainted the back of the case in a brown color. I found a partial manual on the Internet as well as a newer roll chart than the one it came with. I tested a number of tubes with it and it appeared to be working fine.

In closing, when using a tube tester of this type, I would suggest a few things to watch out for:

1. For best accuracy, adjust the **SET LINE** switch again after inserting the tube as it can load down the circuit.

2. Turn the filament voltage to zero after testing a tube. If you forget to do this, Murphy's law will ensure that you will plug in an expensive 1 volt tube after using it to test a 50 volt tube and immediately burn out the filament!

3. Move all levers back to the center position before testing a new tube, or you may apply a voltage that can damage the tube.

4. Some tubes have multiple lines on the roll chart for testing multiple elements. Don't forget to perform all the tests.

Remember that, particularly with emission type testers, the merit test is only a rough indicator of a good tube. A tube that tests as bad may work fine in a real circuit and a tube that tests as good may not work.

And finally, if you get strange results, it is possible that there is an error in the roll chart. You can compare the settings for different revisions of the roll chart and see how they compare. Also compare the result to another tube of the same type, if available.

Chapter 11: Miscellaneous Test Equipment

In this section we cover some miscellaneous types of test equipment that did not fit in any of the previous categories.

Signal tracers are a type of electronic test equipment used for radio and amplifier servicing. When using a signal tracer, the service person would trace a signal from the antenna through each stage of the radio in turn until reaching the speaker, or in the case of an inoperative radio, the faulty stage. The other common method of diagnosing radio faults was to use a signal generator to inject signals, starting with the final audio stage and working back to the antenna until the fault was found. Signal tracers tended to be lower cost devices than signal generators, but by the 1960s signal generators were preferred as the instrument to use for troubleshooting, in part because they were also needed for alignment of RF and IF radio stages. Heathkit made several models of signal tracers. Most units could detect AF and RF (or IF) signals and output them to a loudspeaker as well as indicate the signal strength on a magic eye tube.

Heathkit offers some specialized instruments used mainly for measuring audio signals, such as those from stereo and hi-fi amplifiers. A couple of instruments, like the AA-1 Audio Analyzer and AW-1 Audio Wattmeter could measure audio power levels, useful for testing amplifiers for rated output. Several models of distortion meters were offered that could measure audio distortion of signals as well as AC voltage levels.

Most televisions that use CRTs make use of a *yoke* for deflecting the electron beam, and a *flyback* circuit for generating the high voltages required. Dedicated yoke/flyback testers like the IT-5235 could test the coil windings on yokes, flybacks, and other types of deflection coils.

Heathkit offered some products specifically for working on digital circuits, those that work with high and low logic levels. The IT-7410 Logic Probe indicated the logic levels of TTL or CMOS digital circuitry as well as pulses, using lights. The ID-4804 Byte Probe could measure the state of up to eight digital inputs. The IC-1001 Logic Analyzer could record activity on up to 16 digital signals and display information about the captured data on a terminal or computer attached through a serial port.

The EU-805A and EU-805D were billed as a Universal Digital Instrument. These were a single instrument which could perform seven functions including counting events, acting as a DMM, a frequency counter and more. The EU-805D version lacked the DMM function and was less expensive than the EU-805A which sold for $1250 in 1971.

An unusual product was the IDW-100, a set of 12 instruments that simulated an X-ray machine and was intended for teaching and training. The IDW-39 was the same system less the oscilloscope. The level of simulation must have been quite crude given the instruments used.

What I would consider the ultimate Heathkit instrument for pure geek appeal is an analog computer like the Heathkit EC-1 Educational Analog Computer (Illustration 39). State of the art in 1961 and weighing in at 48 pounds, it allowed one to solve mathematical problems (like calculating the "flight of a projectile") by connecting different types of circuits together via patch cords and reading the result on a meter. Heathkit also offered the similar H-1 and ES-400 analog computers. Needless to say, these computers predated the personal computer and did not actually run software as such, they were configured entirely with switches, knobs, and patch cables.

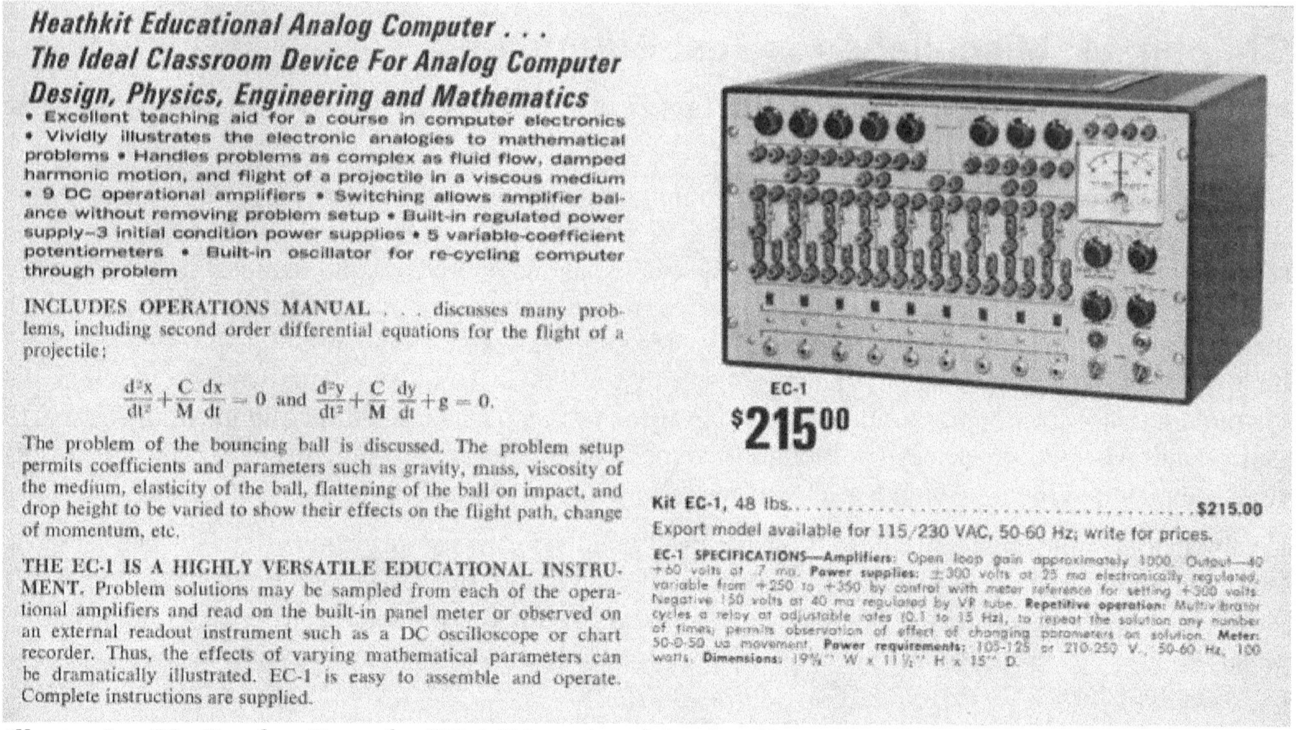

Heathkit Educational Analog Computer . . .
The Ideal Classroom Device For Analog Computer
Design, Physics, Engineering and Mathematics

• Excellent teaching aid for a course in computer electronics • Vividly illustrates the electronic analogies to mathematical problems • Handles problems as complex as fluid flow, damped harmonic motion, and flight of a projectile in a viscous medium • 9 DC operational amplifiers • Switching allows amplifier balance without removing problem setup • Built-in regulated power supply—3 initial condition power supplies • 5 variable-coefficient potentiometers • Built-in oscillator for re-cycling computer through problem

INCLUDES OPERATIONS MANUAL . . . discusses many problems, including second order differential equations for the flight of a projectile:

$$\frac{d^2x}{dt^2} + \frac{C}{M}\frac{dx}{dt} = 0 \text{ and } \frac{d^2y}{dt^2} + \frac{C}{M}\frac{dy}{dt} + g = 0.$$

The problem of the bouncing ball is discussed. The problem setup permits coefficients and parameters such as gravity, mass, viscosity of the medium, elasticity of the ball, flattening of the ball on impact, and drop height to be varied to show their effects on the flight path, change of momentum, etc.

THE EC-1 IS A HIGHLY VERSATILE EDUCATIONAL INSTRUMENT. Problem solutions may be sampled from each of the operational amplifiers and read on the built-in panel meter or observed on an external readout instrument such as a DC oscilloscope or chart recorder. Thus, the effects of varying mathematical parameters can be dramatically illustrated. EC-1 is easy to assemble and operate. Complete instructions are supplied.

EC-1
$215⁰⁰

Kit EC-1, 48 lbs. $215.00
Export model available for 115/230 VAC, 50-60 Hz; write for prices.

EC-1 SPECIFICATIONS—Amplifiers: Open loop gain approximately 1000. Output—40 +60 volts at .7 ma. Power supplies: ±300 volts at 25 ma electronically regulated, variable from +250 to +350 by control with meter reference for setting +300 volts. Negative 150 volts at 40 ma regulated by VR tube. Repetitive operation: Multivibrator cycles a relay at adjustable rates (0.1 to 15 Hz), to repeat the solution any number of times, permits observation of effect of changing parameters on solution. Meter: 50-0-50 ua movement. Power requirements: 105-125 or 210-250 V., 50-60 Hz, 100 watts. Dimensions: 19⅜" W x 11½" H x 15" D.

Illustration 39: Catalog Entry for EC-1 Educational Analog Computer

Heathkit typically sold all the tools necessary for building their kits as well as other equipment useful for working on electronics. These were not manufactured or branded by Heathkit but were approved and endorsed by them. Items in this category include solder, soldering irons, wire strippers, pliers and screwdrivers. They also sold several types of lamps including magnifying lamps and various types of vises and circuit board holders. These were generally of good quality and being able to pick them up at your local Heathkit store, or by mail order, was convenient for many kit builders.

Finally, a Heathkit that probably fell the farthest outside the scope of electronic equipment might be the Boonie-Bike, shown in Illustration 40. This was a small motorcycle that could be driven off-road. Small enough to fit in a car trunk, it could still be driven by an adult. It even came with optional lights and a ski for driving it on snow! This was just one example of how Heathkit could think "outside the box" and offer products that were popular and unique, even if they did not fit into their traditional product lines.

Go Places...On The Heathkit "Boonie-Bike"...The All-Season Trail And Snow Bike

Kit GT-18
$199⁹⁵
(less ski, horn, lights & special muffler)

Discover The Exciting World Of The Boondocks . . . With The Amazing Heathkit "Boonie-Bike"

• More versatile than other trail bikes at twice the price • Rugged, reliable transportation over virtually any kind of terrain • Famous Briggs & Stratton 5 Hp, 4 cycle engine • 2 speed, chain drive transmission • Automatic centrifugal clutch • Hand operated Bendix rear brake for greater stopping ability • Welded tubular steel frame • Front and rear steel fenders for safe, dry riding • Twist grip throttle • Spring loaded front suspension • Tubeless tires—giant 18x8.50x8 rear tire for amazing traction • Extra large cushioned seat • Optional Ski Accessory quickly converts the bike for use on snow • Add the optional horn & lights kit for even greater versatility • Semi-kit for easy one-evening assembly with the famous Heathkit manual

Welcome To The World Of The Boondocks . . remote backwoods areas . . . rough country roads . . . isolated lakes and streams . . . and a world more. It's all waiting for you when you have the unique Heathkit GT-18. It's larger & huskier than a mini-bike . . . smaller, lighter & substantially more powerful than a motorcycle-type trail bike . . . and it has the agility, stability, traction and sheer guts of a mountain goat. It's the new "Boonie-Bike" and it's the only way to go when the going gets rough.

Solid, Heavy-Duty Construction. Mounted on the rigid tubular steel frame is a 5 HP, 4-cycle Briggs & Stratton engine that gives extraordinary power to the 116 lb. "Boonie-Bike". The 3-quart tank lets you run up to 60 miles before refueling, and 6¾" of ground clearance and a welded steel skid pan give worry-free running over territory that would ruin many other bikes.

The Widest Wheel In Trail Bikes. Both tires come premounted & inflated and are top quality name brand tubeless types. The giant rear tire is the big secret behind the GT-18's fantastic performance . . . it's a whopping 18 x 8.50 x 8" . . . and it delivers real traction in mud, sand, gravel, snow, tall weeds and rough underbrush. For even greater versatility, add the ski that snaps on in seconds without tools and lets you go in snow . . . and the new horn & lights and recharging kits that permit night use.

Go! A flip of the shift lever mounted in front of the large, comfortable seat and you're on the way. Select low gear for hills and rough country, high for relatively flat terrain and speeds up to 30 mph. The big 5" Bendix drum-type rear brake is there when you need it for safe, sure stops.

Family Fun. Take the GT-18 camping, hunting . . . to the summer cottage or mountain retreat. It fits easily in car trunks. So easy to ride anyone can master it with a few minutes instruction. So much fun you'll be looking for reasons to ride it. Order yours now . . . and go places.

Kit GT-18, Trail Bike, 140 lbs., no money dn....................................$199.95
Model GTA-18-1, Ski Accessory, 12 lbs.......................................$19.95
Kit GTA-18-2, Horn & Lights Accessory, 10 lbs., (less battery).......$29.95
Model GTA-18-3, Battery Charger Accessory, 3 lbs.............................$6.95

GT-18 SPECIFICATIONS—Engine: Briggs & Stratton 5 h.p. 4 cycle. **Clutch:** Centrifugal. **Crankcase oil capacity:** 1¼ pt. **Transmission:** 2 speed. **Weight:** Net 116 lbs. **Speed range:** Hi 0-30 mph; Lo 0-15 mph. **Gear Ratio:** Lo Speed 15 to 1; Hi Speed 7.5 to 1. **Load capacity:** 400 lbs. **Gas tank capacity:** 3 Qt. (U.S.) **Gas consumption:** Approx. 2 hrs. per tank (approx. 60 miles). Approx. 80 mpg. **Brake:** Bendix 5" internal expanding. **Front suspension:** Spring loaded front suspension. **Rear tire:** Size 18x8.50-8; Tubeless 18" diam., 8.5 width, 8" rim. **Front tire size:** 5.30x4.50-6. **Color:** Blue & White. **Dimensions: Length (overall):** 59", with ski, 70". **Width (at handlebars):** 24½". **Hgt.:** (at handlebars): 35½", (at seat): 25½". **Wheel base:** 42". **Ground clearance:** 6¾"

New Horn & Lights Accessory Kit Makes The "Boonie-Bike" Even More Versatile

The new Horn & Lights Kit is the fast, low cost way to add the extra safety & convenience to your GT-18 that night travel demands.

A powerful 5¾" sealed-beam headlamp in a chrome housing has both High & Low beams for safety, and On/Off and High/Low switches are mounted on top of the lamp for easy accessibility. The 3¾" taillight/stoplight in a chrome housing means maximum visibility and uses a standard automotive bulb for easy replacement. The 4" chrome plated horn is triggered by a button on the handlebars. All cables are held neatly in place by a spiral wrap. The battery holder matches the GT-18, & details in the manual aid in procuring the recommended battery locally. An inexpensive Charger Kit is also available to keep your battery in peak operating condition always.

Kit GTA-18-2, Horn & Lights Accessory,
10 lbs. (less battery) **$29.95**
Assembled GTA-18-3 Battery Charger,
3 lbs...................................**$6.95**

New Spark-Arresting Muffler For Safe Travel In Wooded Areas

For those planning to travel in forests and wooded areas, Heath now offers this U.S. Department of Agriculture approved Spark-Arresting Muffler. Specifically designed for 5 HP Briggs & Stratton engines, the muffler is chrome plated & easy to install. Play it safe . . . don't take a chance on starting a forest fire . . . order yours now.

GTA-18-4, 4 lbs...................**$9.95**

Illustration 40: Catalog Page for the Boonie-Bike

Table 17: Heathkit Miscellaneous Test Equipment

Model	Description	First Year	Last Year	Comments
AA-1	Audio Analyzer	1958		Generator, AC VTVM, wattmeter, distortion meter
AF-1	Audio Frequency Meter	1951	1959	20 Hz to 100 kHz
AW-1	Audio Wattmeter	1951	1960	5 mW to 50 W, 10 Hz to 250 kHz
BT-1	Battery Tester	1951	1960	0 to 15 V and 0 to 180 V, 10 ma or 100 ma load, for testing A or B batteries
EC-1	Analog Computer			Educational, 48 lbs
EK-1	Basic Electricity Course	1960	1967	Includes VOM
EK-2A	Basic Radio Course			Includes regenerative receiver
EK-2B	Basic Radio Course			Expands EK-2A to superhet receiver
EK-3	Basic Transistors Course			Includes intercom
ES-200	Amplifier	1956	1956	For use with ES-400 computer
ES-400	Analog Computer	1956	1963	Up to 15 amplifiers, 30 coefficient plus two aux 10 turns potentiometers, 6 initial conditions floating power supplies, 2 relays, 1 clock oscillator, 1 reference power supply, 1 metering circuit
ETI-7050	Storage Cabinet	1990	1990	For ETI series
EU-51A	Modular Breadboard System			EU-53A plus parts and drawer
EU-53A	Modular Breadboard System			Solderless breadboard system
EU-80	Voltage Reference	1970	1970	Accurate DC voltage standard
EU-80A	Voltage Reference	1970	1970	Accurate DC voltage standard
EU-200-01	Potentiometric Amplifier	1971	1972	Chemical lab instrument
EU-805A	Universal Digital Instrument	1970	1970	With DMM
EU-805D	Universal Digital Instrument	1970	1970	Without DMM
EUW-16	Voltage Reference	1964		0 to 100V, 0 to 90 mV in 10 mV Steps. Ranges X1, X10, X100, X1000
EUW-19A	Operational Amplifier	1963	1966	Analog computer
EUW-19B	Operational Amplifier	1967	1967	Analog computer
EVA-20-12	PH/MV Test Unit	1964	1970	Chemical lab instrument
H-1	Analog Computer	1956		70 tubes, 15 amplifiers
IC-1001	Logic Analyzer	1987	1991	16 inputs, connects to terminal or computer over serial port
ID-4804	Byte Probe	1987	1987	8 inputs
ID-5252	Audio Load	1977	1979	2 to 32 Ω, 60 to 240 W, mono or stereo
IDS-100	X-Ray Machine Simulator			Includes Oscilloscope
IDW-39	X-Ray Machine Simulator			Less Oscilloscope
IM-20	Battery Tester	1961	1963	Replaced BT-1
IT-12	Signal Tracer	1963	1977	RF probe, speaker, eye tube
IT-5235	Yoke/Flyback Tester	1979	1983	For television servicing
IT-5283	Signal Tracer	1977	1991	RF probe, speaker, solid-state

Model	Description	First Year	Last Year	Comments
IT-7400	Digital IC Tester			14 and 16 pin devices, TTL, RTL, ECL, etc.
IT-7410	Logic Probe	1978	1987	Indicates logic level or pulses
SM-2376	Insulation Tester	1989	1992	Clamp-on
ST-2204	Telephone Line Analyzer	1987	1987	No details known
ST-5235	Yoke/Flyback Tester			Assembled version of IT-5235
T-1	Signal Tracer			Speaker
T-2	Signal Tracer	1950	1951	Speaker
T-3	Signal Tracer	1951	1957	Speaker
T-4	Signal Tracer	1958	1962	Magic eye indicator
TO-1	Test Oscillator	1959		5 frequencies: 2.62, 455, 465, 600, 1400 kHz or 2 external crystals, 400 Hz modulation.
VT-1	Vibrator Tester	1953	1959	Used in power supplies

In-Depth: EK-2B Radio Receiver and Course

For an entry in the miscellaneous section, we'll look at the EK-2B Radio Receiver and course. While ostensibly a radio, it was a little different from the other products in that it was sold as part of an educational training course on electronics and radio theory. The course, actually a series of courses, included both theory and hands-on experiments, which culminated in building the radio.

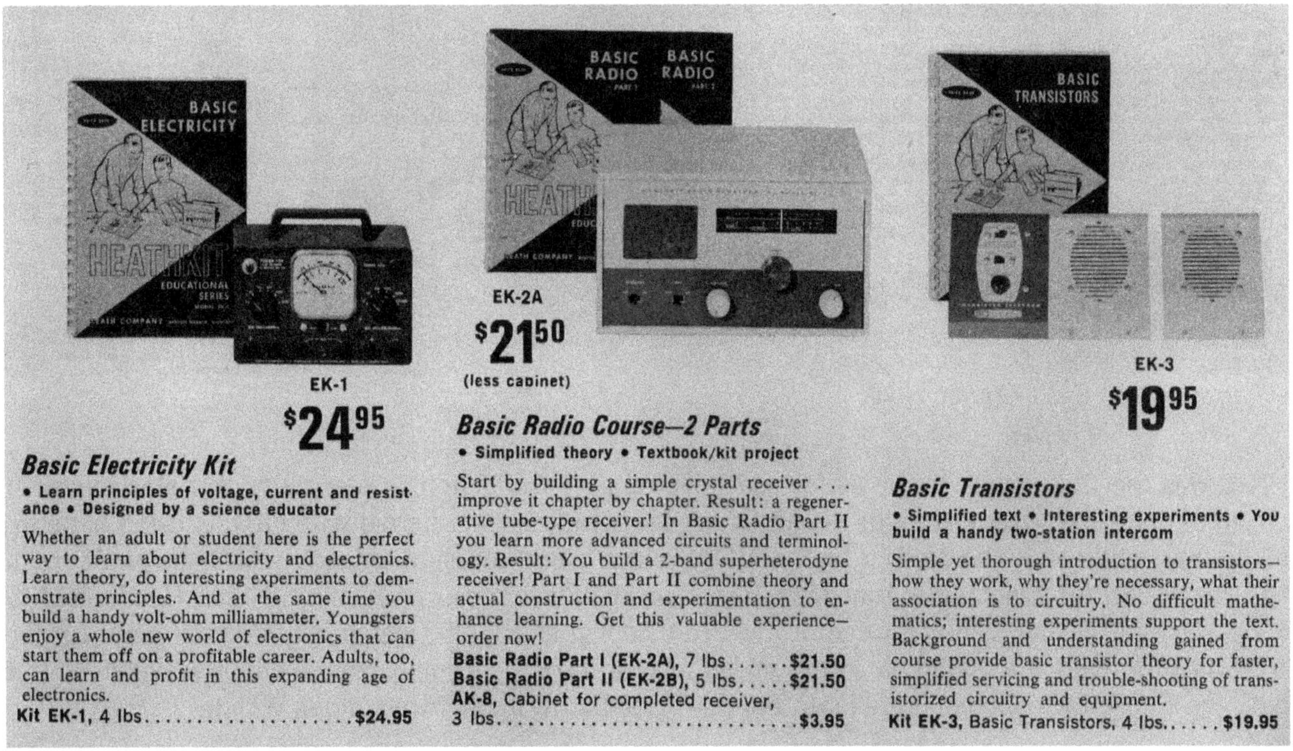

Illustration 41: 1971 Catalog Entry for EK Series Courses

It was part of a series of four courses (Illustration 41): EK-1, EK-2A and EK-2B, and EK-3. All of the courses were offered separately and each consisted of a workbook with about 100 pages of training material as well as the electronic components needed for the hands-on labs. Unlike the Heathkit assembly manuals that came with kits, these were complete courses that included theory, hands-on experiments, and exercises.

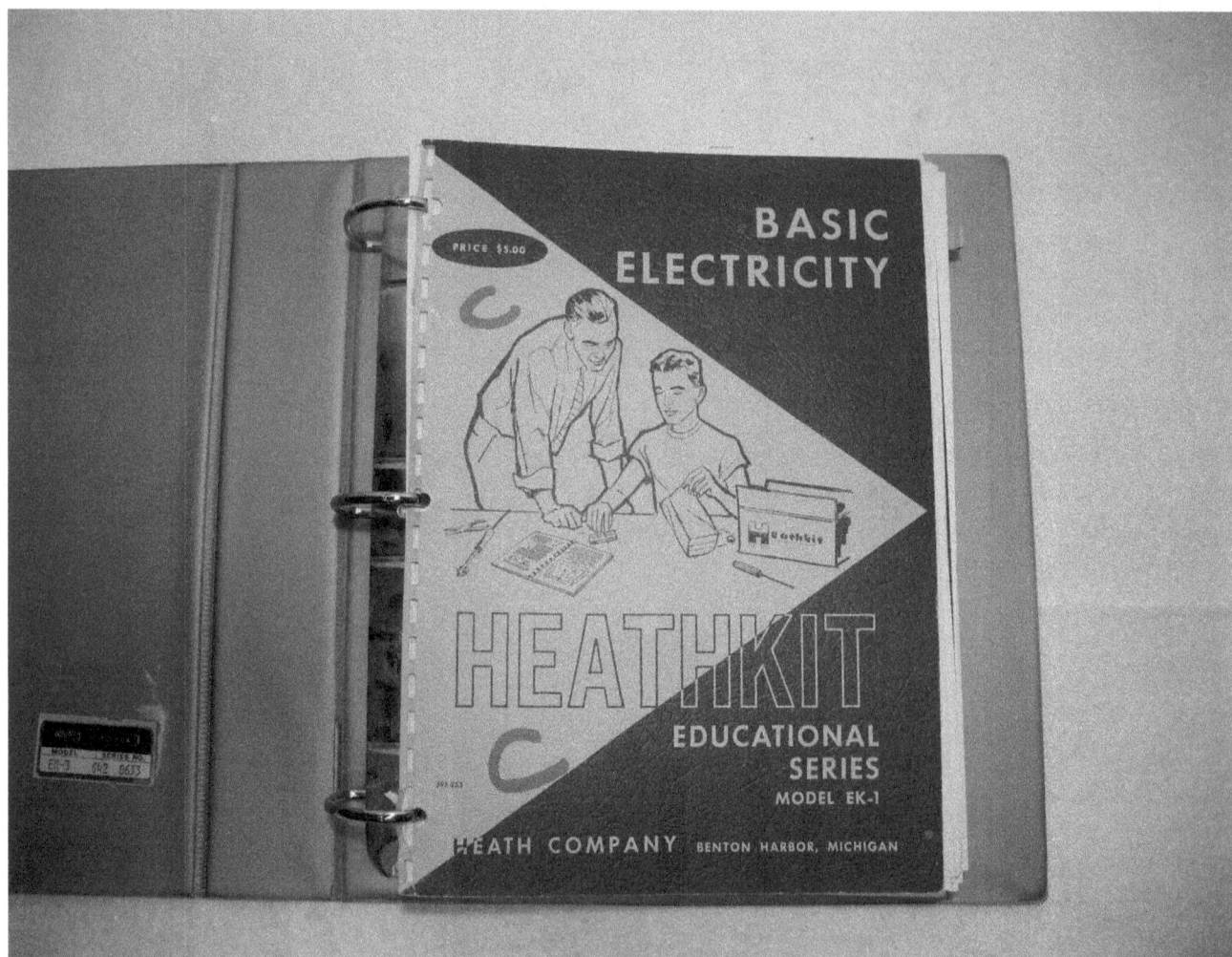

Illustration 42: EK-1 Course Book

EK-1 was the first course in the series and covered an introduction to electricity and electronics. It covered topics like what electricity is and Ohm's law. Experiments started with circuits using batteries and lamps. Ultimately the student built a VOM which could measure voltage, resistance and current and was useful for general electronics work such as appliance repair and building other Heathkits. My 1961 Heathkit catalog lists the price of the EK-1 course at $27.95 (which included the parts for the VOM) and in 1971 it was $24.95. I have an original of the EK-1 course manual which is dated 1959. I have the course material but, regrettably, not the VOM hardware that went along with it.

Illustration 43: EK-2A Course Book

Courses EK-2A and EK-2B covered basic radio theory and built on the fundamentals from EK-1. EK-2A covered basic radio concepts and built up a two tube regenerative receiver. My manual is dated December 1960. Course EK-2B continued with a more detailed presentation of radio theory that built up a 6-tube 2-band superheterodyne receiver, the EK-2B that I have. My manual is dated December 1964. In my 1961 catalog the EK-2A and EK-2B courses were selling for $29.95 each. For an additional $5.95 you could buy the AK-8 cabinet for the receiver, otherwise it only consisted of an open chassis.

The EK-2B radio that the student eventually built up, a stage at a time, offered coverage of two bands: the AM broadcast band from 540 to 1600 kHz and shortwave from 3 to 10 MHz. It uses a 455 kHz IF frequency and sported an adjustable BFO so you could receive Morse code and single sideband transmissions as well as AM. It utilized six tubes, and was basically a design called the *All American Five*, the standard five tube radio design that was common from the late 1940s into the early 1960s, with an additional tube to support the BFO, as well as band switching to add a shortwave band. The radio did not have a built in antenna. It needed an external one although a short piece of wire would suffice to pick up local AM broadcasts and a few shortwave stations.

The performance is not great, particularly on shortwave, but the purpose was really to learn, and having a working radio to show for your efforts would have been very satisfying.

EK-3 was a course on basic transistors, and was introduced a little later than the EK-2A and 2B courses that focused on vacuum tubes. It covered the basic theory of how transistors work and resulted in building a little two station intercom system.

Illustration 44: EK-3 Course Book

My copy of the course material is dated December 1961. This was early days for transistors, and shows how Heathkit was on the forefront of technology at the time. The kit uses two 2N1274 transistors which mounted on sockets. These are germanium transistors, which were common at the time but today most transistors and integrated circuits use silicon. Incidentally, my Heathkit HW-16 amateur radio transceiver uses one 2N1274 transistor in the receiver circuit (the rest of the design uses tubes). The EK-3 course is not listed in my 1961 Heathkit catalog but in my 1971 catalog it was selling for $19.95.

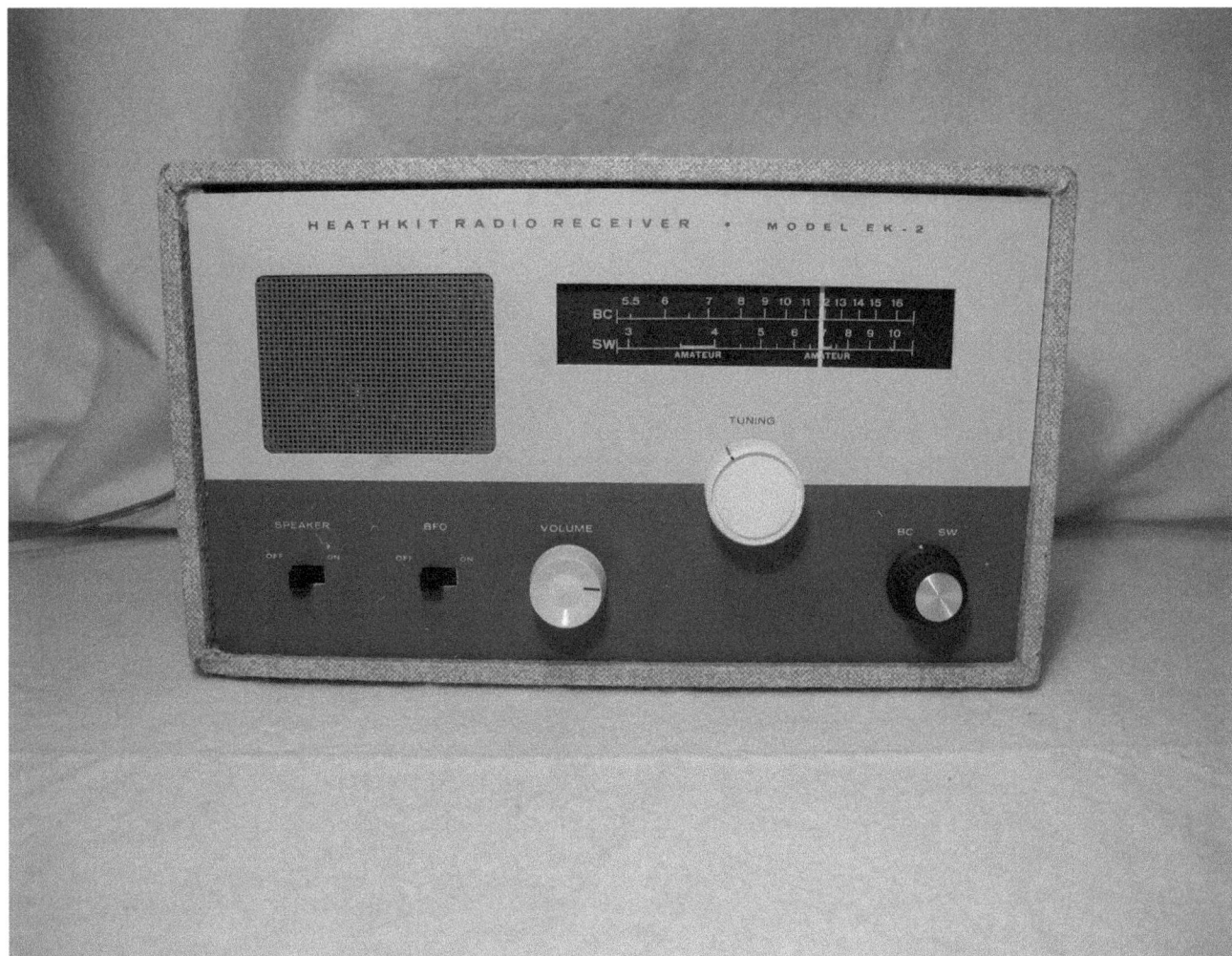

Illustration 45: EK-2B Front View

Illustration 45 shows the completed EK-2B with the optional case. The front panel has switches for BFO and speaker, an on/off and volume control, band switch, and tuning. Tuning uses a pretty standard slide rule dial with dial cord arrangement. Being able to switch the speaker independently of the headphones is a little unusual; most designs used a headphone jack that muted the speaker when headphones were plugged in. On the back are antenna and ground lugs and the headphone jack.

Illustration 46: EK-2B Inside View

Removing the case you can see the parts are built on a standard metal chassis using point to point wiring. Most of the wiring is under the chassis. The original builder (not me) did a good job of putting this unit together.

Note that the radio uses a power transformer. Most of the AA5 type radios at the time omitted this, which saved money but typically meant that the chassis had live AC power on it, which was extremely dangerous. A power transformer isolates the unit from AC power making it safer although today I doubt that a line operated radio with exposed high voltage like this would be targeted as being suitable for young people to build.

I got my unit of this radio in September 2005 on eBay. It came with the optional case, which had some damage but was repairable - it is made of thin wood with a paper covering. The 5Y3GT rectifier tube was intermittent and had a visibly broken wire to the filament. I was able to pick up some stations on the AM band. Two of the original knobs were missing. I replaced the 5Y3GT tube, and put on some different knobs. I did a full alignment of the radio. It did not have any further circuitry or component problems and works quite well.

Prepare for the "Scientific 60's... with the Heathkit® Educational Series

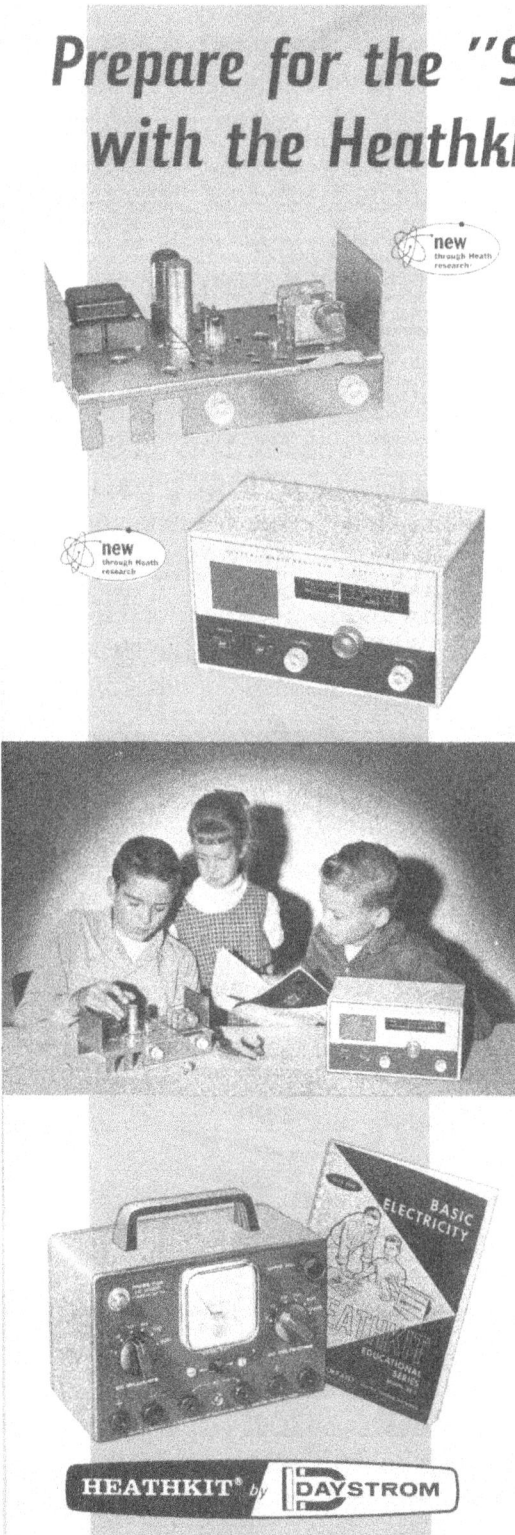

A NEW TWO PART COURSE IN BASIC RADIO . . .
EDUCATIONAL KITS EK-2A (part I) and EK-2B (part II)

Here's a new 2-part Kit and Text-workbook project designed to teach practical, basic radio theory to everyone. Functions and concepts are described in everyday language and illustrated by common analogies that eliminate difficult mathematics. You start with the EK-2A, "Basic Radio—Part I," learning the basic parts of a radio. Typical chapters are, "How Does Voice and Music Get From the Broadcast Station to Your Home?"—"What Is a Detector?"—"What is a Tuned Circuit?" With this groundwork you build a crystal receiver and, chapter by chapter, improve upon it by adding parts until you have a regenerative tube-type receiver which pulls in stations hundreds of miles away. The EK-2B or "Basic Radio—Part II" advances your radio theory knowledge with such subjects as "What is a Local Oscillator?"—"What is a Mixer?"—"What Does Alignment Mean?", etc. With the additional parts supplied, you improve upon the receiver from Part I until at the completion of Part II you have a 2-band superheterodyne receiver that picks up broadcast, amateur radio, marine and international stations. Designed for youngsters and adults alike, the EK-2 series provides an excellent medium for learning basic radio at home or in classrooms. The wealth of this series is not in the finished product, a fine little 6-tube 2-band superheterodyne receiver, but in the complete and authoritative information contained in the manuals themselves and in the integration and continuity that has been achieved—between theory and practical experiments. Start now on an exciting and challenging career in electronics by exposing yourself (or your son) to basic electronic knowledge. The continuing Heath Educational series is designed to explain electronics through a learn-by-doing process . . . providing, at the same time, practical instruments to enhance the learning process.

Kit Model EK-2A . . . 6 lbs. . . . (less cab.)		**$29.95**
Kit Model EK-2B . . . 4 lbs. . . . (less cab.)		**$29.95**
Accessory Cabinet AK-8: A handsome housing for your completed EK-2B Receiver . . . 4 lbs.		**$5.95**

teaches basic "yardsticks" of electronics . . .

EDUCATIONAL KIT—EK-1

This priceless introduction into the wonderful world of electronics opens up fascinating areas of study for youngsters and adults alike. The EK-1 is a combination kit and text-workbook designed to teach and demonstrate the principles of voltage, current and resistance, the theory and construction of series and parallel direct current circuits, voltmeter, ammeter and ohmmeter circuits and the application of Ohm's law to these circuits. The completed meter is used to verify Ohm's law and the maximum power transfer theorem, one of the most important theorems in electronics. The finished kit, a practical *DC volt-ohm-milliammeter*, may be used in a wide variety of applications. Procedures for checking automobile circuits, home appliances, etc., are included in the manual. The EK-1 will provide an excellent background for the EK-2 and future Heathkit Educational kits, so get started now in this fascinating learn-by-doing series. 4 lbs.

Kit Model EK-1	**$27.95**

Illustration 47: Catalog Entry for EK-1 and EK-2 Courses

My unit did not come with a manual. I purchased a full set of the EK-1 through EK-3 manuals on eBay. They make for interesting reading and I read through most of the material.

The courses were aimed at both adults and young people. You can see a little blatant sexism of the time in the catalog copy from 1961 which says "Start now on an exciting and challenging career in electronics by exposing you (or your son) to basic electronics". Pictured are two young boys working on the radio while their sister watches on in the background. I would think that the full series of courses could take several months to complete if the student diligently followed all of the material, did the labs, and wrote the quizzes.

These were early training courses. At the time Heathkit also offered courses on test equipment like oscilloscopes and VTVMs that included test equipment kits. In the later days of Heathkit, training courses were the major part of their business, continuing after they exited the kit business in the 1990s.

For its time, this series of courses were an excellent way to get started in electronics, either as a hobbyist or for one who wanted to make a career out of it. Keep in mind that in the 1950s and 60s, radio and television repair was a viable business. Despite not being a particularly good receiver, unbuilt versions of this kit have sold for over $600 on eBay, capitalizing on the nostalgia factor.

Postscript: The Ubiquitous Heathkit Nut Starter

I'd like to conclude the main portion of the book with a short tribute to one of the Heathkit innovations that almost every kit builder will remember with nostalgia. A nut starter is a tool for installing threaded nuts in hard to reach areas. Without one, when installing nuts, screws, and washers in a chassis it often seems that you need three hands and it can take several attempts, with you possibly dropping parts inside the chassis or on the floor in the process. Almost every Heathkit included a little red plastic nut starter which proved indispensable, not just when building the kit, but as a general tool to be used again and again in the workshop. A simple plastic cylinder, it had different size openings on each end so it could fit various size nuts or screws (e.g. #2 through #6). With the insertion of a piece of flat metal (included in the kit if applicable) it could also be used an alignment tool for adjusting coils and transformers.

USING A PLASTIC NUT STARTER

A plastic nut starter offers a convenient method of starting most sizes: 3/16'' and 1/4'' (3-48 and 6-32). When the correct end is pushed down over a nut, the pliable tool conforms to the shape of the nut and the nut is gently held while it is being picked up and started on the screw. The tool should only be used to start the nut.

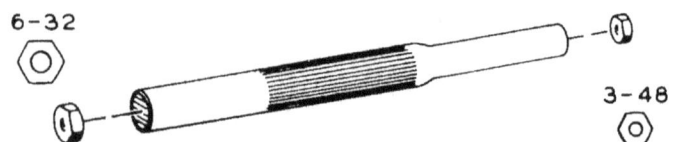

6-32

3-48

Heathkit gave it part number 490-5 and, as well as being included in kits where it was considered useful for assembly, it could be purchased as an individual replacement part. A 1967 kit manual lists the cost at 15 cents. Used and completed Heathkits rarely include the nut starter but I imagine most builders must have kept them in their tool drawer after the kit was built. If you get your hands on one I encourage you to hold onto it. If you don't have one, several companies sell nut starters ranging from plastic reproductions very similar to Heathkit's to much more expensive steel units. I've seen people assembling modern amateur radio kits struggling to assemble the hardware, being advised to use a Heathkit nut starter or find a modern reproduction.

Appendix A: References and Resources

This sections lists books, web sites, and other resources of interest to collectors of Heathkit test equipment and vintage electronics.

Books

Here are some published books related to Heathkit and test equipment.

Heathkit - A Guide to the Amateur Radio Products by Chuck Penson (2nd Edition, May 2003)

This book describes all of the amateur radio products offered by Heathkit as well as a history of Heathkit with an emphasis on their amateur radio products. It also include tips on collecting and restoring.

Heathkit: The Early Years by Terry A. Perdue (Jun 2001)

This is a CD-ROM collection of over 1,000 images of Heathkit-related photographs, catalogs, flyers, and other documents as well as a 30 minute audio recording by Heathkit Director of Engineering Gene Fiebich. (As far as I can determine this is no longer available.)

Heath Nostalgia by Terry A. Perdue (Jan 1992)

This book, by a former Heath engineer, tells the story of Heathkit and the people who worked there. (The book is out of print and no longer available except for used copies.)

Tube Testers and Classic Electronic Test Gear by Alan Douglas (Aug 2000)

This book covers tube-based test equipment and their manufacturers. Tube testers are covered in detail, as well as VOMs, VTVMs, grid dip meters, and other test instruments. It includes coverage of some Heathkit equipment.

Web Sites and Other Internet Resources

Web sites come and go regularly on the Internet so these are all subject to change, but here are a few that I have found useful.

http://ebaman.com/

This site has a collection of manuals for vintage electronics, including Heathkit. It originally started as a mirror of the now defunct site BAMA (Boat Anchor Manual Archive) and was subsequently expanded in scope.

http://en.wikipedia.org/wiki/Heathkit

Wikipedia has a good entry covering the history of the Heathkit company and their major product lines. It also has entries describing most of the major types of electronic test equipment.

https://puck.nether.net/mailman/listinfo/heath

This site hosts a large and active mailing list for owners and collectors of Heathkit equipment of all types.

http://groups.yahoo.com/group/heathkit/

Yahoo! hosts a Heathkit user group and associated mailing list for Heathkit products including test equipment. It currently has over 2600 members including some former Heathkit employees.

http://heathkit-museum.com/

This site, The Heathkit Virtual Museum, has a lot of interesting information about Heathkit including a section on test equipment.

http://www.d8apro.com/

This is the site of Data Professionals who purchased the rights to Heathkit's legacy product documentation. They sell copies of manuals for almost all kits produced by Heathkit.

http://www.ebay.com/bhp/heathkit-test-equipment

This is the eBay category for Heathkit test equipment. At any given time there is probably on the order of 500 items listed, making this the single largest source of Heathkit test equipment in the world.

http://www.heathkit.com/

This is the official Heathkit web site, until recently the home of Heathkit Educational Systems, and still up at the time of writing.

http://www.heathkit.nu/

Despite the name, an unofficial Heathkit fan site created by Swedish Heathkit collector Hans Gatu with a detailed company history, details on a number of pieces of Heathkit test equipment and amateur radios in his collection, and links to other sites.

http://www.mods.dk/

This site has a large collection of free Heathkit manuals, ranging from schematics to partial and full manuals. The site is run by volunteers and donations are welcomed which will give you improved access to the site.

http://www.nostalgickitscentral.com/

This site has a lot of information about classic electronic kits including Heathkit. There are pictures, schematics, and specs for hundreds of Heathkit test equipment models.

http://www.radiomuseum.org/

This site hosted in Germany is focused on vintage radio but has information on a lot of test equipment as well. Gaining full access requires either making a donation or submitting material to the site.

http://www.rsp-italy.it/Electronics/Kits/

This site has a large collection of images, schematics, and manuals for electronic kits including Heathkit.

http://www.vintage-radio.info/heathkit/

This site hosts the Heathkit Schematic Diagram Archive which includes information on several hundred models of Heathkit equipment.

http://www.w6ze.org/Heathkit/Heathkit_Index.html

Bob Eckweiler, amateur radio call sign AF6C, writes a monthly *Heathkit of the Month* column for the Orange Country Amateur Radio club newsletter. His articles often cover test equipment and are archived at this site.

http://vintagemanuals.com

Vintage Manuals sell copies of vintage radio, audio and test equipment manuals, including most models manufactured by Heathkit.

http://tubularelectronics.com/

This site has a collection of scans of old radio and electronics books and manuals including older Heathkit manuals for which copyrights are believed to be expired.

http://justradios.com/

JustRadios sells vintage radio schematics, capacitors and resistors including high voltage electrolytic capacitors that can be hard to find from the normal parts suppliers, and "cap kits" of replacement capacitors for specific radio models.

http://www.youtube.com/user/jefftranter

Since 2011 I have been making YouTube videos demonstrating vintage radio and test equipment, much of it from Heathkit. As of the time of writing I have made over 40 videos and have more than 300 YouTube subscribers with more than 100,000 views. Most of the Heathkit models featured in the In-Depth sections of this book have corresponding videos that you may wish to view.

Appendix B: Product Listing

I've attempted here to present a comprehensive list of all the Heathkit test equipment that was manufactured. This information came from several sources and undoubtedly contains some errors and omissions. Heathkit produced over 400 unique models of test equipment over the years, too many to cover each one individually in the book. Models described in the book are listed in the index.

Table 18: Heathkit Test Equipment Sorted by Model

Model	Description	First Year	Last Year	Comments
225	Adaptor			TV Picture Tube Test Adaptor for tube checker
309	Probe - RF	1952		For use with V-1 VTVM
310	Probe - HV			10 kV
336	Probe - HV	1952		30 kV for use with VTVMs
342	Probe - Low Capacitance			X1, Oscilloscope
355	Adaptor			TV Picture Tube Test Adaptor for use with TC-2
309-B	Probe - RF			RF up to 250 MHz, for use with V-1
309-C	Probe - RF			For use with VTVM, scope, signal tracer
337-B	Probe - RF Demodulator			For use with VTVM, scope, signal tracer
337-C	Probe - RF Demodulator			For oscilloscope for signal tracer
338-B	Probe - Peak-to-Peak			5 Hz to 5 MHz
AA-1	Audio Analyzer	1958		Generator, AC VTVM, Wattmeter, distortion meter
AD-1309	Noise Generator	1985	1991	White and pink noise generator
AF-1	Audio Frequency Meter	1951	1959	20 Hz to 100 kHz
AG-7	Signal Generator - AF	1951	1951	20 Hz to 200 kHz, sine/square wave
AG-8	Signal Generator - AF	1951	1957	18 Hz to 1 MHz, sine wave
AG-9	Signal Generator - AF	1956	1957	20 Hz to 110 kHz, sine wave
AG-9A	Signal Generator - AF			10 Hz to 110 kHz, sine wave
AG-10	Signal Generator - AF	1958	1962	20 Hz to 1 MHz, sine/square wave
AO-1	Signal Generator - AF	1951	1959	20 Hz to 20 kHz, sine/square wave
AV-1	VTVM	1951	1951	10 ranges
AV-2	VTVM - AC	1951	1956	0.01V RMS to 300V RMS
AV-3	VTVM - AC	1957	1961	10 ranges
AW-1	Audio Wattmeter	1951	1960	5 mW to 50 W, 10 Hz to 250 kHz
BG-1	Bar Generator	1953	1956	Horizontal and vertical bars for B&W TV
BT-1	Battery Tester	1951	1960	0 to 15 V and 0 to 180 V, 10 ma or 100 ma load, for testing A or B batteries
C-1	Capacitor Checker	1948	1950	Magic eye indicator
C-2	Capacitor Checker	1950	1951	Magic eye indicator
C-3	Capacitor Checker	1951	1960	Magic eye indicator
CC-1	Tube Tester - CRT	1956	1960	Emission type
CD-1	Color Bar/Dot Generator	1957	1962	For television servicing

Model	Description	First Year	Last Year	Comments
CM-1	Capacitance Meter	1956	1960	Analog meter
CS-1	Capacitor Substitution Box	1953	1962	18 values from .0001 to 0.22µF
CT-1	Capacitor Checker	1957	1960	Magic eye, opens and shorts test only
DC-1	Capacitor Substitution Box	1951	1961	Decade box, 3 knobs
DR-1	Resistor Substitution Box	1956	1961	Decade box, 5 knobs
EC-1	Analog Computer			Educational, 48 lbs
EF-1	VTVM Course			For IM-18, IM-28, etc. Includes power supply.
EF-2	Oscilloscope Course			Includes test chassis. EF-2-3 includes IO-21. EF-2-5 includes IO-18
EF-3	Signal Generator Course			Includes test chassis. EF-3-2 includes IG-102
EK-1	Basic Electricity Course	1960	1967	Includes VOM
EK-2A	Basic Radio Course			Includes regenerative receiver
EK-2B	Basic Radio Course			Expands EK-2A to superhet receiver
EK-3	Basic Transistors Course			Includes intercom
ES-1	Power Supply	1956	1956	For use with ES-400 computer
ES-100	Power Supply	1956	1956	For use with ES-400 computer
ES-200	Amplifier	1956	1956	For use with ES-400 computer
ES-400	Analog Computer	1956	1963	Up to 15 amplifiers, 30 coefficient plus two aux 10 turns potentiometers, 6 initial conditions floating power supplies, 2 relays, 1 clock oscillator, 1 reference power supply, 1 metering circuit
ES-600	Function Generator	1956	1956	For ES-400 computer
ETI-7010	DMM	1990	1990	Bench type, assembled
ETI-7020	Function Generator	1990	1990	With counter, assembled
ETI-7030	Power Supply	1990	1990	Bench type, triple output
ETI-7040	Frequency Counter	1990	1990	8 digit, factory assembled, Heath Zenith
ETI-7050	Storage Cabinet	1990	1990	For ETI series
EU-20A	Chart Recorder	1968	1970	Malmstadt-Enke Instruments series
EU-20M	Chart Recorder	1968	1970	Malmstadt-Enke Instruments series
EU-30A	Resistor Substitution Box	1970	1970	Decade box, 7 knobs
EU-40	Power Supply	1969	1970	0 to 300V @ 20mA, 6.3 VAC @ 1A
EU-40A	Power Supply			50 to 300V @20mA, 6.3VAC @1A
EU-41	Power Supply	1969	1970	0 to 15V @ 750mA
EU-41A	Power Supply			0 to 15V @750mA
EU-51A	Modular Breadboard System			EU-53A plus parts and drawer
EU-53A	Modular Breadboard System			Solderless breadboard system
EU-70	Oscilloscope	1970	1972	15 MHz, dual trace, solid-state, assembled
EU-80	Voltage Reference	1970	1970	Accurate DC voltage standard
EU-80A	Voltage Reference	1970	1970	Accurate DC voltage standard

Appendix B: Product Listing

Model	Description	First Year	Last Year	Comments
EU-81	Function Generator	1971	1972	0.1 Hz to 1 MHz
EU-200-01	Potentiometric Amplifier	1971	1972	Chemical lab instrument
EU-805A	Universal Digital Instrument	1970	1970	With DMM
EU-805D	Universal Digital Instrument	1970	1970	Without DMM
EUP-26	VOM	1968	1970	Weston Meter
EUW-15	Power Supply	1964	1970	200 to 350 VDC @ 100mA, 6.3VAC @ 3A
EUW-16	Voltage Reference	1964		0 to 100V, 0 to 90 mV in 10 mV Steps. Ranges X1, X10, X100, X1000
EUW-17	Power Supply	1967	1970	0 to 25V @ 200mA
EUW-18	Lab Meter	1967	1970	1 mA meter with shunts
EUW-19A	Operational Amplifier	1963	1966	Analog computer
EUW-19B	Operational Amplifier	1967	1967	Analog computer
EUW-20A	Chart Recorder	1963	1970	Malmstadt-Enke Instruments series, wired
EUW-20M	Chart Recorder	1966	1967	Malmstadt-Enke Instruments series, wired
EUW-24	VTVM	1963	1970	4.5", 7 ranges, assembled
EUW-25	Oscilloscope	1963	1970	3", 400 kHz
EUW-27	Signal Generator - AF	1968	1970	20Hz to 1MHz, sine/square, assembled
EUW-28	Resistor Substitution Box	1967	1970	36 values from 15Ω to 10 MΩ
EUW-29	Capacitor Substitution Box	1967	1970	18 values from .0001 to 0.22µF
EUW-30	Resistor Substitution Box	1968		Same as IN-17
EUW-301	Chart Recorder	1964	1966	Malmstadt-Enke Instruments series, wired
EUW-301M	Chart Recorder	1966	1967	Malmstadt-Enke Instruments series, wired
EV-3	Oscilloscope	1964	1970	IMPScope biological EKG type
EVA-20-12	PH/MV Test Unit	1964	1970	Chemical lab instrument
EVW-3	Oscilloscope	1968		Assembled version of EV-3
FMO-1	FM Sweep Alignment Generator	1960	1967	90/100/107 MHz, 400Hz modulation, 10.7MHz sweep
G-1	Signal Generator - RF	1950	1950	150 kHz to 102 MHz
G-2	Signal Generator - Sine/Square			20 Hz to 20 kHz, sine/square wave
G-3	Sweep Generator	1948	1949	FM 10.7 MHz
G-4	Signal Generator - RF			150 kHz to 30 MHz
G-5	Signal Generator - RF	1950	1950	150 kHz to 102 MHz
GD-1	Grid Dip Meter	1951	1951	With plug-in coils
GD-1A	Grid Dip Meter	1953	1953	With plug-in coils
GD-1B	Grid Dip Meter	1953	1960	With plug-in coils
H-1	Analog Computer	1956		70 tubes, 15 amplifiers
HD-1	Harmonic Distortion Meter	1958		Measures voltage and % distortion

Classic Heathkit Electronic Test Equipment   122

Model	Description	First Year	Last Year	Comments
HD-1250	Dip Meter	1975	1991	Solid state, 1.6 MHz to 250 MHz, 7 plug-in coils
HM-10	Tunnel Dipper	1961	1962	Solid state (tunnel diode), plug-in coils
HM-10A	Tunnel Dipper	1962	1970	Solid state (tunnel diode), plug-in coils
IB-1	Impedance Bridge	1950	1950	Battery powered, wood case
IB-1B	Impedance Bridge	1951	1966	Battery powered, wood case
IB-2	Impedance Bridge	1951	1956	Measures R, L, C, D, Q. Built in 1 kHz generator
IB-2A	Impedance Bridge	1957	1967	Measures R, L, C, D, Q. Built in 1 kHz generator
IB-28	Impedance Bridge	1968	1976	Measures R, L, C, D, Q. Built in 1 kHz generator
IB-101	Frequency Counter	1970	1972	5 digit, 15MHz, Nixie display
IB-102	Frequency Scaler	1971	1975	Extends range of IB-101
IB-1100	Frequency Counter	1973	1975	5 digit, 30MHz, Nixie display
IB-1101	Frequency Counter	1972	1975	5 digit, 100MHz, Nixie display
IB-1102	Frequency Counter	1972	1977	8 digit, 120 MHz, Nixie display
IB-1103	Frequency Counter	1973	1977	8 digit, 180 MHz, Nixie display
IB-3128	Impedance Bridge	1977	1980	Same as IB-28 but different paint
IB-5281	Impedance Bridge	1977	1990	Measures R, L, C. Solid-state, battery powered.
IC-62	Color Generator			For television servicing
IC-1001	Logic Analyzer	1987	1991	16 inputs, connects to terminal or computer over serial port
ID-22	Electronic Switch	1964	1970	Converts scope to dual trace
ID-101	Electronic Switch	1971	1976	Converts scope to dual trace
ID-2311	DMM	1989	1991	3.5 digit
ID-4101	Electronic Switch	1977	1981	Converts scope to dual trace
ID-4804	Byte Probe	1987	1987	8 inputs
ID-4850	Digital Memory Oscilloscope	1989	1992	Digital memory box for scopes
ID-5252	Audio Load	1977	1979	2 to 32 Ω, 60 to 240 W, mono or stereo
IDS-100	X-Ray Machine Simulator			Includes Oscilloscope
IDW-39	X-Ray Machine Simulator			Less Oscilloscope
IF-1272	Signal Generator - AF	1977	1982	Low distortion
IG-14	Marker Generator	1968	1970	Sine/square
IG-18	Signal Generator - AF	1969	1975	Sine/square
IG-18A	Signal Generator - AF	1975	1977	Sine/square
IG-28	Color Generator	1969	1977	For television servicing, produces 12 patterns
IG-37	FM Stereo Generator	1968	1976	For alignment of FM radios
IG-42	Signal Generator - RF	1962	1979	Replaced LG-1
IG-52	TV Alignment Generator	1967	1972	For television servicing
IG-57	Post Marker/Sweep Generator	1968	1971	For television servicing
IG-57A	Post Marker/Sweep Generator	1970	1978	For television servicing

Model	Description	First Year	Last Year	Comments
IG-62	Bar/Dot Generator	1967	1968	For television servicing
IG-72	Signal Generator - AF	1962	1977	10 Hz to 100 kHz
IG-82	Signal Generator - AF	1963	1968	Sine/square
IG-102	Signal Generator - RF	1963	1977	100 kHz to 110 MHz
IG-102S	Signal Generator - RF			Berkeley Physics Laboratory version of IG-102
IG-112	FM Stereo Generator	1964	1967	For alignment of FM radios
IG-1271	Function Generator	1975	1987	Sine/square/triangle, 0.1 Hz to 1 MHz
IG-1272	Signal Generator - AF	1977	1983	5 Hz to 100 kHz, low noise
IG-1273	Function Generator	1977	1979	Sine/square/triangle, 0.3 Hz to 3 MHz
IG-1275	Function Generator	1977	1983	Sine/square/triangle, 0.3 Hz to 3 MHz
IG-1277	Pulse Generator	1984	1987	Variable pulse period, width, delay
IG-4244	Oscilloscope Calibrator	1983	1992	For calibrating oscilloscopes
IG-4505	Oscilloscope Calibrator	1975	1990	For calibrating oscilloscopes
IG-5218	Signal Generator - AF	1977	1990	Sine/square, to 100 kHz
IG-5228	Bar/Dot Generator	1977	1983	For television servicing, 12 patterns
IG-5237	FM Stereo Generator	1977	1979	For alignment of FM radios
IG-5240	Color Generator	1976	1984	Hand-held, battery operated, 16 patterns
IG-5242	Signal Generator - RF	1979	1979	Same as IG-42 but with "new look"
IG-5257	Post Marker/Sweep Generator	1977	1984	For television servicing
IG-5260	TV Alignment Generator	1989	1991	Hand-held, battery operated
IG-5280	Signal Generator - RF			310 kHz to 110 MHz
IG-5282	Signal Generator - AF	1977	1991	Sine/square, 10 Hz to 100 kHz
IGW-18	VTVM			Berkeley Physics Laboratory, assembled
IGW-19	Signal Generator - RF			Berkeley Physics Laboratory, assembled version of IG-102S
IGW-47B	Signal Generator - Sine/Square			Berkeley Physics Laboratory, assembled, same as EUW-27
IGW-57A	Post Marker/Sweep Generator	1970	1978	Assembled version of IG-57A
IM-1	Distortion Analyzer	1951	1954	Measures IMD, full scale ranges of 30%, 10%, 3%
IM-10	VTVM	1960	1962	Deluxe model, 6" meter, 7 ranges
IM-11	VTVM	1961	1968	Replaced V-7A
IM-12	Harmonic Distortion Meter	1963	1967	Analog meter
IM-13	VTVM	1963	1968	Bench type, 6" meter
IM-16	VOM	1967	1974	Bench type, 6" meter, solid-state
IM-17	VOM	1967	1977	High impedance FET, battery-powered, in storage case
IM-18	VTVM	1968	1976	4.5", 7 ranges, probe
IM-20	Battery Tester	1961	1963	Replaced BT-1
IM-21	VTVM - AC	1961	1968	AC RMS voltage, 0.01 to 300 VAC in 10 ranges

Appendix B: Product Listing

Model	Description	First Year	Last Year	Comments
IM-22	Audio Intermodulation Analyzer	1962	1968	Replaced AA-1
IM-25	VOM	1967	1974	Bench type, 6" meter, solid-state, high-impedance
IM-28	VTVM	1968	1976	Bench type, 6" meter
IM-30	Transistor Tester	1961	1967	Tests bipolar transistors for shorts, gain, leakage
IM-32	VTVM	1962	1963	Replaced IM-10
IM-36	Transistor Tester	1968	1974	Tests bipolar transistors for shorts, gain, leakage
IM-38	VTVM - AC	1968	1976	AC RMS voltage, 0.01 to 300 VAC in 10 ranges
IM-48	Audio Intermodulation Analyzer	1968	1976	AC VTVM, Wattmeter, distortion analyzer
IM-58	Harmonic Distortion Meter	1969	1976	Measures noise and distortion, 20 to 20 kHz
IM-102	DMM	1971	1978	3.5 digit, Nixie display, 26 ranges
IM-103	Line Voltage Monitor	1970	1976	90 to 140 VAC
IM-104	VOM	1973	1976	FET VOM, 4.5" meter
IM-105	VOM	1971	1981	Taut band 4.5" meter
IM-1104	VOM	1979	1981	10 MΩ input, portable
IM-1202	DMM	1973	1975	2.5 digit, Nixie display
IM-1210	DMM	1977	1981	2.5 digit, LED display
IM-1212	DMM	1975	1976	2.5 digit, Nixie display
IM-2202	DMM	1975	1981	3.5 digit, LED display
IM-2212	DMM	1979	1981	3.5 digit, LED display, auto-ranging
IM-2215	DMM	1979	1984	3.5 digit, LCD display, hand-held
IM-2260	DMM	1982	1991	3.5 digit, LED display
IM-2262	DMM	1982	1983	3.5 digit, LCD display, true RMS
IM-2264	DMM	1982	1987	3.5 digit, LCD display, true RMS, crest factor indicator
IM-2320	DMM	1987	1990	3.5 digit, LCD display, hand-held
IM-2400	Frequency Counter	1980	1991	7 digit, 512 MHz, hand-held
IM-2410	Frequency Counter	1980	1992	8 digit, 225 MHz
IM-2420	Frequency Counter	1980	1990	8 digit, 512 MHz, two inputs
IM-4100	Frequency Counter	1975	1979	5 digit, 30 MHz, period and totalize
IM-4110	Frequency Counter	1977	1979	8 digit, 110 MHz
IM-4120	Frequency Counter	1977	1979	8 digit, 250 MHz
IM-4130	Frequency Counter	1977	1979	8 digit, 1 GHz
IM-4180	FM Deviation Meter	1979	1987	For FM transmitter testing
IM-4190	Bi-Directional Wattmeter	1978	1981	Forward and reflected power, 100 to 1000 MHz, 300W
IM-5210	Probe Meter	1975	1984	40 kV HV probe meter
IM-5215	Probe Meter	1984	1991	40 kV HV probe meter
IM-5217	VOM	1979	1987	Comes with case
IM-5218	VTVM	1977	1983	4.5", 7 ranges, assembled
IM-5225	Multimeter	1977	1981	FET analog

Model	Description	First Year	Last Year	Comments
IM-5228	VTVM	1977	1989	Bench type version of IM-5218, 6" meter
IM-5238	VTVM - AC	1976	1981	1 mV to 300 V AC, 12 ranges
IM-5248	Distortion Analyzer	1976	1979	Distortion 0.1 to 100% in 3 ranges, AC voltmeter 10 mV to 100 V
IM-5258	Distortion Analyzer	1976	1983	Distortion 0.03 to 100% in 6 ranges, AC voltmeter 1 mV to 300 V
IM-5284	VOM	1977	1983	Solid-state, battery or AC power
IMA-18-1	Solid State Tube Replacements			Replaces tubes to convert IM-18, IM-28 and other VTVMs to solid state
IMA-100-10	Probe - HV		1987	X100 30 kV HV probe for 10 MΩ meters
IMA-100-11	Probe - HV	1985	1986	X100 30 kV HV probe for 11 MΩ meters
IMA-1000-1	Probe - HV			X1000 HV probe for 1 MΩ meters
IMA-2215-1	Capacitance Meter	1981	1987	Assembled version of IT-2250
IN-11	Resistor Substitution Box	1961	1967	Decade box, 6 knobs
IN-12	Resistor Substitution Box	1962	1967	36 values from 15Ω to 10 MΩ
IN-17	Resistor Substitution Box	1967	1978	Decade box, 6 knobs
IN-21	Capacitor Substitution Box	1961	1967	Decade box, 3 knobs
IN-22	Capacitor Substitution Box	1962	1967	18 values from .0001μF to 0.22μF
IN-27	Capacitor Substitution Box	1967	1977	Decade box, 3 knobs
IN-37	Resistor Substitution Box	1967	1978	36 values from 15Ω to 10 MΩ
IN-47	Capacitor Substitution Box	1967	1978	18 values from .0001μF to 0.22μF
IN-3117	Resistor Substitution Box	1991		Decade box, 6 knobs
IN-3127	Capacitor Substitution Box	1981		Decade box, 3 knobs
IN-3137	Resistor Substitution Box	1977	1981	36 values from 15Ω to 10 MΩ
IN-3147	Capacitor Substitution Box	1977	1981	18 values from .0001μF to 0.22μF
IO-10	Oscilloscope	1960	1967	3", 200 kHz, recurrent sweep
IO-12	Oscilloscope	1962	1968	5", 4 MHz
IO-14	Oscilloscope	1966	1971	5", 8 MHz
IO-17	Oscilloscope	1968	1973	3", 5 MHz
IO-18	Oscilloscope	1968	1970	5", 5 MHz
IO-21	Oscilloscope	1961	1972	3", 200 kHz
IO-30	Oscilloscope	1960	1962	5", 5 MHz
IO-101	Vectorscope/Color Generator	1970	1977	3" vectorscope and color bar/pattern generator
IO-102	Oscilloscope	1971	1975	5", 5 MHz
IO-103	Oscilloscope	1972	1974	5", 10 MHz
IO-104	Oscilloscope	1973	1975	5", 15 MHz
IO-105	Oscilloscope	1971	1974	5", 15 MHz, dual trace
IO-1128	Oscilloscope	1971	1974	3", vector monitor
IO-3220	Oscilloscope	1982	1984	5", 20 MHz, dual trace, battery powered
IO-4101	Oscilloscope	1977	1980	Vectorscope like IO-101

Model	Description	First Year	Last Year	Comments
IO-4105	Oscilloscope	1979	1987	5", 5 MHz
IO-4205	Oscilloscope	1979	1987	5", 5 MHz, dual trace
IO-4210	Oscilloscope	1989	1990	5", 10 MHz, dual trace
IO-4225	Oscilloscope	1989	1990	5", 25 MHz, dual trace
IO-4235	Oscilloscope	1979	1984	5", 35 MHz, dual trace, delayed sweep
IO-4360	Oscilloscope	1984	1984	5", 60 MHz, triple trace
IO-4510	Oscilloscope	1974	1980	5", 15 MHz, dual trace
IO-4530	Oscilloscope	1975	1977	5", 10 MHz, TV Service
IO-4540	Oscilloscope	1975	1976	5", 5 MHz, hobby/service
IO-4541	Oscilloscope	1977	1979	5", 5 MHz, special TV triggering
IO-4550	Oscilloscope	1976	1984	5", 10 MHz, dual trace
IO-4555	Oscilloscope	1978	1979	5", 10 MHz
IO-4560	Oscilloscope	1975	1979	5", 5 MHz, auto triggered sweep
IOA-3220-1	Probes - Oscilloscope			Two PKW-105 probes with pouch
IOA-4200	Oscilloscope Timebase Module	1984	1986	Digital display for IO-4360 and IO-4225
IOA-4510-1	Oscilloscope Calibrator			For IO-4510 or IO-4530
IOW-18S	Oscilloscope			Berkeley Physics Laboratory, 5" laboratory
IP-10	Isolation Transformer	1961	1962	90 to 130 VAC, 300W, meter
IP-12	Power Supply	1962	1975	Battery Eliminator. 6V unfiltered @10A, filtered @5A or 12V unfiltered @ 5A, filtered @5A
IP-17	Power Supply	1968	1977	0 to 40V @100mA, 0 to -100V @1mA, 6.3VAC @4A, 12.6VAC @2A
IP-18	Power Supply	1968	1977	1 to 15V @ 500mA
IP-20	Power Supply	1962	1967	0 to 50V @ 1.5A
IP-22	Isolation Transformer	1963	1964	90 to 130VAC, 300W, meter
IP-27	Power Supply	1968	1975	0.5 to 50V @1.5A, meter
IP-28	Power Supply	1969	1976	1 to 10V or 1 to 30V @1A, meter
IP-32	Power Supply	1962	1967	0 to 100V @ 1mA, 0 to 400V @ 100mA
IP-2670	Power Supply	1975		7.5 to 15V, meters
IP-2700	Power Supply	1975	1976	0 to 60V @ 1.5A, analog meters
IP-2701	Power Supply	1975	1977	0 to 60V @ 1.5A, digital meters
IP-2710	Power Supply	1975	1981	0 to 30V @ 3A, analog meters
IP-2711	Power Supply	1975	1981	0 to 30V @ 3A, digital meters
IP-2715	Power Supply	1975	1983	Battery eliminator. 9-15V@12A, meters
IP-2717	Power Supply	1977	1982	0 to 400 V @ 125mA, 0 to 100V @ 1mA, 6.3VAC, 12.6VAC
IP-2717A	Power Supply	1982	1991	0 to 400V @ 100mA, 0 to 100V@1mA. 6.3VAC@4A, 12.6VAC @ 2A, meters
IP-2718	Power Supply	1976	1992	5V @1.5A, 2 X 0 to 20V @0.5A, meter
IP-2720	Power Supply	1975	1977	0 to 15V @ 5A, analog meters

Model	Description	First Year	Last Year	Comments
IP-2721	Power Supply	1975	1976	0 to 15V @ 5A, digital meters
IP-2728	Power Supply	1977	1990	1 to 15V @ 500mA
IP-2730	Power Supply	1975	1977	0 to 7.5V @ 10A, analog meters
IP-2731	Power Supply	1975	1977	0 to 7.5V @ 10A, digital meters
IP-2760	Power Supply	1984	1987	Battery eliminator - restyled IP-2715
IP-5220	Power Supply	1975	1983	0 to 140VAC with isolation
IPA-5280-1	Power Supply	1977	1989	Power supply for IT-5280 series instruments
IPW-17	Power Supply			Berkeley Physics Laboratory
IPW-27	Power Supply	1968	1975	Assembled version of IP-27
IR-18M	Chart Recorder	1971	1979	Two input ranges, 12 speeds
IR-5204	Chart Recorder	1980	1983	Standard chart recorder
IR-5207	Plotter X-Y	1979	1981	Chart recorder with X-Y mode
IS-1	Isolation Transformer			Two voltage settings, meter
IT-1	Isolation Transformer	1953	1960	Two voltage settings, meter
IT-10	Transistor Checker	1961	1967	Tests bipolar transistors for shorts, gain, leakage
IT-11	Capacitor Checker	1961	1968	Magic eye indicator
IT-12	Signal Tracer	1963	1977	RF probe, speaker, eye tube
IT-17	Tube Tester	1967	1976	Emission type, shorts, leakage,opens, filament test
IT-18	Transistor Tester	1968	1978	Tests bipolar transistors for shorts, gain, leakage
IT-21	Tube Tester	1961	1967	Replaced TC-3
IT-22	Capacitor Checker	1963	1967	Magic eye, opens and shorts test only
IT-27	Transistor Checker	1967	1978	Tests bipolar transistors for shorts, gain, leakage
IT-28	Capacitor Checker	1968	1977	Magic eye indicator
IT-121	Transistor Tester	1973	1976	Tests bipolar, FETs, SCRs, triacs, etc. Gain and transconductance.
IT-1121	Curve Tracer	1974	1977	Tests transistors, diodes. Used with scope.
IT-2127	Transistor Tester	1978	1981	Same as IT-37
IT-2232	Component Tracer	1984	1990	Built in CRT
IT-2240	Impedance Bridge	1989	1990	Digital benchtop
IT-2250	Capacitance Meter	1981	1987	Digital hand-held
IT-3117	Tube Tester	1977	1981	Same as IT-17
IT-3118	Transistor Tester	1977	1979	Same as IT-18
IT-3120	Transistor Tester	1977	1989	Tests bipolar, FETs, SCRs, triacs, etc. Gain and transconductance.
IT-3121	Curve Tracer	1978	1983	Tests transistors, diodes. Used with scope.
IT-3127	Transistor/Diode Checker	1978	1983	Simple diode and transistor tester
IT-5230	Tube Tester	1976	1990	CRT tester and rejuvenators
IT-5235	Yoke/Flyback Tester	1979	1983	For television servicing
IT-5283	Signal Tracer	1977	1991	RF probe, speaker, solid-state
IT-7400	Digital IC Tester			14 and 16 pin devices, TTL, RTL, ECL, etc.

Appendix B: Product Listing

Model	Description	First Year	Last Year	Comments
IT-7410	Logic Probe	1978	1987	Indicates logic level or pulses
ITA-5230-1	Adaptor	1976	1990	CRT socket adaptor for IT-5230
LG-1	Signal Generator - RF	1953	1962	150 kHz to 30 MHz
LP-1	Linearity Pattern Generator	1955	1956	For television servicing
LP-2	Linearity Pattern Generator	1957	1957	For television servicing
M-1	VOM	1950	1962	Pocket-sized "Handitester"
MM-1	VOM	1951	1967	5", 20KΩ per volt, Battery operated portable
O-1	Oscilloscope	1947	1948	5"
O-2	Oscilloscope	1948		5"
O-3	Oscilloscope	1948	1948	5", 150 kHz
O-4	Oscilloscope	1949	1949	5", 2 MHz
O-5	Oscilloscope	1950	1950	5", 2.2 MHz
O-6	Oscilloscope	1950	1951	5", 200 kHz
O-7	Oscilloscope	1951	1951	5", 250 kHz
O-8	Oscilloscope	1951	1953	5", 2 MHz
O-9	Oscilloscope	1951	1954	5", 3 MHz
O-10	Oscilloscope	1955	1956	5", 400 kHz, PC board
O-11	Oscilloscope	1957	1957	5", 5 MHz
O-12	Oscilloscope	1958	1990	5", 5 MHz
OL-1	Oscilloscope	1955	1956	5", 5 MHz
OM-1	Oscilloscope	1955	1956	5", 5 MHz
OM-2	Oscilloscope	1957	1957	5", First Heathkit product
OM-3	Oscilloscope	1958	1960	5", 1.2 MHz
OP-1	Oscilloscope	1958	1962	2.2 MHz
OR-1	Oscilloscope	1959	1962	5", 200 kHz
PK-1	Probe - Low Capacitance			X1 X10 for IO-14 or IO-18
PK-3	Probe - RF			100 MHz, for VTVM
PK-3A	Probe - RF			For DC voltmeters
PKW-2	Probe - Oscilloscope			X1 X10 25 MHz
PKW-4	Probe - for IM-5218/IM-5228			VTVM replacement probe
PKW-101	Probe - Oscilloscope			X10 60 MHz
PKW-104	Probe - Oscilloscope			X1 17 MHz
PKW-105	Probe - Oscilloscope			X1 15 MHz, X10 80 MHz
PKW-200	Test Lead Set	1981		Shielded red/black cables with probes
PM-1	RF Power Meter			100 kHz to 250 MHz
PS-1	Power Supply	1950	1951	50 to 300V
PS-2	Power Supply	1951	1954	160 to 450V, 6.3 VAC @ 4A
PS-3	Power Supply	1957		500V @ 200mA, 6.3VAC
PS-4	Power Supply	1961		0 to 400V @ 100 mA, 0 to -100 V @ 1 mA, 6.3VAC @ 4

Model	Description	First Year	Last Year	Comments
				A, meters
QM-1	Q Meter	1951	1964	Similar to other impedance bridges. Measures L, C, Q.
RD-1	Resistor Substitution Box	1950	1951	Decade box, 5 knobs
RF-1	Signal Generator - RF	1960	1962	100 kHz to 110 MHz
RS-1	Resistor Substitution Box	1951	1962	36 values from 15Ω to 10 MΩ
S-1	Electronic Switch	1950	1950	Dual trace for scope
S-2	Electronic Switch	1950	1955	Dual trace for scope
S-3	Electronic Switch	1956	1962	Dual trace for scope
SDS-5000	Oscilloscope	1988	1988	Computer-based
SG-1	Signal Generator - RF			Square wave
SG-5	Signal Generator - RF	1950	1950	160 kHz to 50 MHz
SG-6	Signal Generator - RF	1951	1951	160 kHz to 50 MHz
SG-7	Signal Generator - RF	1951	1953	150 kHz to 150 MHz
SG-8	Signal Generator - RF	1956	1961	160 kHz to 110 MHz
SG-18A	Signal Generator - Sine/Square			Assembled version of IG-18
SG-57A	Post Marker/Sweep Generator	1970	1978	Assembled version of IG-57A
SG-1271	Function Generator			Assembled version of IG-1271
SG-1272	Signal Generator - AF			Assembled version of IG-1272
SG-1274	Function/Pulse Generator	1988	1992	2 MHz swept, assembled
SG-5218	Signal Generator - AF			Assembled version of IG-5218
SM-20A	VTVM			Assembled version of IM-18
SM-21A	VTVM			Bench type, assembled
SM-22A	VTVM - AC			Assembled version of IM-38
SM-104	Frequency Counter			No details known
SM-105	Frequency Counter	1971	1971	80 MHz
SM-105A	Frequency Counter			80 MHz
SM-600	Digital VTVM			No details known
SM-660	VOM	1971	1981	Assembled version of IM-105
SM-666	VOM	1973	1976	Assembled version of IM-104
SM-1210	DMM			3 digit, LED
SM-1212	DMM	1975	1976	Assembled version of IM-1212
SM-2206	DMM	1983	1986	Clamp-on multimeter
SM-2208	DMM	1988	1990	Clamp-on ammeter, 300A
SM-2215	DMM	1979	1984	Assembled version of IM-2215
SM-2255	DMM	1988	1990	3.5 digit, hand-held
SM-2260	DMM	1982	1991	Assembled version of IM-2260
SM-2300	DMM	1986	1988	Shirt pocket sized, assembled
SM-2300A	DMM	1989	1991	Shirt pocket sized, assembled

Appendix B: Product Listing

Model	Description	First Year	Last Year	Comments
SM-2310A	DMM	1989	1989	Shirt pocket sized, assembled
SM-2311	DMM	1989	1992	Assembled version of ID-2311
SM-2320	DMM	1986	1986	Heavy duty portable, assembled
SM-2360	DMM	1989	1990	Bench/portable, assembled
SM-2372	DMM/Frequency Meter	1989	1992	20MHz, assembled
SM-2374	DMM	1989	1992	Clamp-on, 1000A, AC/DC volts
SM-2376	Insulation Tester	1989	1992	Clamp-on
SM-2380	DMM	1990	1992	Assembled, Heath branded
SM-2410	Frequency Counter			Assembled version of IM-2410
SM-2420	Frequency Counter			Assembled version of IM-2420
SM-2440	Frequency Counter	1989	1990	Assembled version of IM-2440
SM-4100	Frequency Counter			Assembled version of IM-4100
SM-4180	Frequency Deviation Meter	1979	1987	Assembled version of IM-4180
SM-4190	Frequency Counter			Assembled version of IM-4190
SM-5210	Probe Meter	1975	1984	Assembled version of IM-5210
SMA-2400-1	Swiveling Telescoping Antenna			For frequency counters like the IM-2400 and IM-2420
SO-29	Oscilloscope	1972	1972	Biological, high gain DC
SO-3220	Oscilloscope			Assembled version of IO-3220
SO-4105	Oscilloscope			Assembled version of IO-4105
SO-4205	Oscilloscope			Assembled version of IO-4205
SO-4221	Oscilloscope		1987	5", 20 MHz, dual trace
SO-4226	Oscilloscope	1987	1989	5", 25 MHz, dual trace
SO-4510	Oscilloscope			Assembled version of IO-4510
SO-4521	Oscilloscope	1987	1989	5", 50 MHz, dual trace
SO-4530	Oscilloscope			Assembled version of IO-4530
SO-4540	Oscilloscope			Assembled version of IO-4540
SO-4550	Oscilloscope			Assembled version of IO-4550
SO-4552	Oscilloscope	1989	1992	5", 25 MHz
SO-4554	Oscilloscope	1989	1992	5", 40 MHz
SP-17A	Power Supply			Assembled version of IP-17
SP-18A	Power Supply			Assembled version of IP-18
SP-2700	Power Supply			Assembled version of IP-2700
SP-2701	Power Supply			Assembled version of IP-2701
SP-2710	Power Supply			Assembled version of IP-2710
SP-2711	Power Supply			Assembled version of IP-2711
SP-2717	Power Supply	1981		Assembled version of IP-2717A
SP-2717A	Power Supply			Assembled version of IP-2717A
SP-2718	Power Supply		1992	Assembled version of IP-2718
SP-2720	Power Supply			Assembled version of IP-2720

Model	Description	First Year	Last Year	Comments
SP-2721	Power Supply			Assembled version of IP-2721
SP-2730	Power Supply			Assembled version of IP-2730
SP-2731	Power Supply			Assembled version of IP-2731
SP-2762	Power Supply	1989	1991	0 to 30V @ 3A, analog meters
SQ-1	Signal Generator - AF	1951	1957	Square wave
ST-2204	Telephone Line Analyzer	1987	1987	No details known
ST-5235	Yoke/Flyback Tester			Assembled version of IT-5235
SU-511-50	50 Ω Termination			For counters, oscilloscopes, DC to 1 GHz
T-1	Signal Tracer			Speaker
T-2	Signal Tracer	1950	1951	Speaker
T-3	Signal Tracer	1951	1957	Speaker
T-4	Signal Tracer	1958	1962	Magic eye indicator
TC-1	Tube Tester	1950	1953	Emission tester
TC-2	Tube Checker	1953	1959	Emission tester, wood cabinet
TC-2P	Tube Checker	1953	1958	Portable version of the TC-2
TC-3	Tube Checker	1959	1962	Emission tester
TO-1	Test Oscillator	1959		5 frequencies: 2.62, 455, 465, 600, 1400 kHz or 2 external crystals, 400 Hz modulation.
TS-1	TV Alignment Generator	1950	1950	For television servicing
TS-2	TV Alignment Generator	1950	1953	For television servicing
TS-3	Sweep Generator	1953	1954	For television servicing, 4 MHz to 220 MHz
TS-4	TV Alignment Generator	1956	1956	For television servicing
TS-4A	TV Alignment Generator	1957	1962	For television servicing
TT-1	Tube Tester	1960	1961	Mutual Conductance tester
TT-1A	Tube Tester	1962	1967	Mutual Conductance tester
V-1	VTVM	1947	1948	5 ranges
V-2	VTVM	1948	1949	6 ranges
V-3	VTVM	1949	1949	Battery-operated, engineering problems, never produced
V-4	VTVM	1950	1950	6 ranges
V-4A	VTVM	1950	1950	6 ranges
V-5	VTVM	1951	1951	6 ranges
V-5A	VTVM	1951	1951	6 ranges
V-6	VTVM	1951	1954	7 ranges
V-7	VTVM	1955	1955	7 ranges, First to use PCB
V-7A	VTVM	1956	1961	4.5" meter, 7 ranges
VC-1	Oscilloscope Calibrator	1951	1951	For calibrating oscilloscopes
VC-2	Oscilloscope Calibrator	1953	1956	For calibrating oscilloscopes
VC-3	Oscilloscope Calibrator	1957	1962	For calibrating oscilloscopes
VT-1	Vibrator Tester	1953	1959	Used in power supplies

Alphabetical Index

Index of Tables

Illustration Index

www.ingramcontent.com/pod-product-compliance
Lightning Source LLC
Chambersburg PA
CBHW062027210326
41519CB00060B/7187